THE IPCC AND THE POLITICS OF WRITING CLIMATE CHANGE

The Intergovernmental Panel on Climate Change (IPCC) is one of the most significant global environmental assessment bodies ever established, providing the most authoritative and influential reports on climate change knowledge. This book examines the history and politics of the organisation and how this shapes its assessment practice and the climate knowledge it produces. Developing a new methodology, this book focuses on the actors, activities and forms of authority shaping the IPCC's constructions of climate change. It describes how social, economic and political dynamics influence all aspects of the organisation and its work. This book contributes to understanding the place of science in politics and politics in science and also offers important insights for designing new knowledge bodies for global environmental agreement-making. It is indispensable for students and researchers in environmental studies, international relations and political science, and science and technology studies. This title is also available as Open Access on Cambridge Core.

HANNAH HUGHES is a senior lecturer in International Politics and Climate Change at Aberystwyth University. Her research is focused on questions of knowledge and power and global asymmetries in how we know and collectively address global environmental degradation. Through her research, collaborations and publications, she hopes to shape central sites and processes of global environmental agreement-making. In 2023, she published another book with Cambridge University Press on *Conducting Research on Global Environmental Agreement-Making* that she co-edited with Alice Vadrot.

THE IPCC AND THE POLITICS OF WRITING CLIMATE CHANGE

HANNAH HUGHES
Aberystwyth University

CAMBRIDGE
UNIVERSITY PRESS

CAMBRIDGE
UNIVERSITY PRESS

Shaftesbury Road, Cambridge CB2 8EA, United Kingdom

One Liberty Plaza, 20th Floor, New York, NY 10006, USA

477 Williamstown Road, Port Melbourne, VIC 3207, Australia

314–321, 3rd Floor, Plot 3, Splendor Forum, Jasola District Centre, New Delhi – 110025, India

103 Penang Road, #05–06/07, Visioncrest Commercial, Singapore 238467

Cambridge University Press is part of Cambridge University Press & Assessment, a department of the University of Cambridge.

We share the University's mission to contribute to society through the pursuit of education, learning and research at the highest international levels of excellence.

www.cambridge.org
Information on this title: www.cambridge.org/9781009341530

DOI: 10.1017/9781009341554

First published 2024

A catalogue record for this publication is available from the British Library

Library of Congress Cataloging-in-Publication Data
Names: Hughes, Hannah, 1980– author.
Title: The IPCC and the politics of writing climate change / Hannah Hughes, Aberystwyth University.
Other titles: Intergovernmental Panel on Climate Change and the politics of writing climate change
Description: New York, NY : Cambridge University Press, 2024. | Includes bibliographical references and index.
Identifiers: LCCN 2023054243 | ISBN 9781009341530 (hardback) | ISBN 9781009341554 (ebook)
Subjects: LCSH: Climatic changes. | Intergovernmental Panel on Climate Change.
Classification: LCC QC903 .H845 2024 | DDC 363.7/056–dc23/eng/20240226
LC record available at https://lccn.loc.gov/2023054243

ISBN 978-1-009-34153-0 Hardback

Additional resources for this publication at www.cambridge.org/hughes

To Mum and Dad

Contents

Acknowledgements

I didn't write this book alone.

The book developed from a United Kingdom Economic and Social Research Council (ESRC) PhD studentship (ES/H010394/1) and the generous supervision I received during four years at the International Politics Department at Aberystwyth University from Andrew Linklater, Mike Williams, Richard Jackson and Kamila Stullerova. It benefited from further research through the ESRC project "The Politics of Science in International Climate Cooperation" (ES/W001373/1). The Open Access version was enabled by a United Kingdom Research Innovation (UKRI) award.

To have Andrew Linklater as a supervisor was to experience something new – to be able to talk about how I saw and understood the world without interruption. He was the most incredible listener. He asked a question and stayed completely still, listening while I responded. I often measure my own supervision abilities and the air space I take next to his, and I still have the art of listening to learn. To be allowed to speak in that way gave me confidence in my ability and I needed that. Mike only supervised me for a short time while I was at Aberystwyth and then as a visiting scholar in Ottawa. However, he saw that I needed books in a way I didn't know myself. He would casually suggest that I go and read a book relevant to what I was puzzling, which would inspire me to think a bit harder. In this respect, I must also thank one of the book reviewers. They suggested several titles to help me bridge from where I was to where I needed to get in terms of the analytical contribution that I was trying to make. Those books were immensely helpful in refining order as the real matter of things.

Turning the PhD into a book has been a long process, and there are people that I want to thank for their generous time and support. Mat Paterson has been an incredible mentor, and I am very grateful for all his support, from quick email checks to attending a book workshop that was kindly supported by the School of Law and Politics at Cardiff University. You are wonderful Mat and Jo, thank you. I am grateful for the feedback from Steven Bernstein at the workshop, for his support of all my project ideas and for his company just hanging out, wandering around and eating stuff. Matthew Eagleton-Pierce, Inanna Hamati-Ataya and Alexander Ruser attended the workshop along with colleagues at Cardiff (Ben Pontin and Dot Kwek), and I am grateful for each question and comment that helped me to consider how I was situating the book, putting things together and why. The book reviewers were also excellent in this regard; I had six sets of mostly supportive comments that gave me practical advice for focusing the book and editing its content.

I want to thank my colleagues in the International Politics Department at Aberystwyth University for their support, including funding a research assistant, Lewis Birley, to help prepare the final manuscript. I really began to struggle to immerse myself in re-writing and keeping on top of everything else, missing meetings, forgetting things I'd promised and even turning up an hour late for a lecture. My colleagues quietly sorted out my muddles and gave me space to repair and carry on. Thank you.

I could not have undertaken this project if participants to the IPCC had not generously given their precious time for interviews, follow-up email and discussions. The authors and bureau members I spoke to had already committed incredible effort to the IPCC, and their support of the project made me look again and think a bit deeper at any simple conclusions I might be tempted to reach. It is easy to become disheartened when your work is rejected and ignored. The fact that you answered my emails, kept your doors open and allowed me access kept me going through the many years of learning and revising and meant that this book had to be finished to repay the debt.

I try to stay conscious of the origins of my thinking and reference those that have informed this. Many scholars have inspired how I think and write in this book, but in the end, this may not be reflected in the final text. In the layered process of writing and updating, editing and rewriting, which has been the practice of writing my PhD into a book, much of the nitty-gritty of learning Bourdieu, his concepts and how others applied him has been lost so that I could hone in on my own understanding and the development of an analytical framework developed for and adapted to a study of the IPCC. There is also the reading and learning since the PhD, the scholars and books that have come to shape the changing nature of my understanding but weren't directly related to this study and as a result may not be captured in the references and citations contained within this book. I apologise for any oversight you as a reader may observe.

The bulk of the PhD was written up in a little house on the corner with José Carli and finalised in a new home, overlooking the town as a family of four. I am sorry to all my family and friends for my absences over the years, when I've told you I can't talk right now or haven't messaged back and disappeared into getting this book into the next stage of its development. Thank you for still being there when I re-emerged. Finally, I want to thank Ron Rathbun. He reminded me that in the end, no one can tell me what I have come to understand. These words came back to me many, many times in the last months as I was tempted to send it out for just another pair of eyes and another final check. It is what it is now. Finished. I hope, somehow, my thoughts will contribute to a cleaner, fairer planet.

Acronyms

IPCC

AR4	Fourth Assessment Report
AR5	Fifth Assessment Report
AR6	Sixth Assessment Report
CA	Contributing Author
CLA	Convening Lead Author
FAR	First Assessment Report
IPCC	Intergovernmental Panel on Climate Change
LA	Lead Author
RE	Review Editor
SAR	Second Assessment Report
SPM	Summary for Policymakers
TAR	Third Assessment Report
TS	Technical Summary
TSU	Technical Support Unit
WG	Working Group
WGI	Working Group I
WGII	Working Group II
WGIII	Working Group III

Other

AGGG	Advisory Group on Greenhouse Gases
AAAS	American Association for the Advancement of Science
CDM	Clean Development Mechanism
COP	Conference of the Parties
CSIRO	Commonwealth Scientific and Industrial Research Organisation
CV	Curriculum Vitae

DEFRA	Department of Environment, Food and Rural Affairs
EIT	Economies in Transition
ENB	Earth Negotiations Bulletin
FOD	First Order Draft
GCM	Global Climate Modelling
GDP	Gross Domestic Product
GEA	Global Environmental Assessment
GHG	Greenhouse Gases
GST	Global Stocktake
IAC	InterAcademy Council
IAM	Integrative Assessment Models
ICSU	International Council for Science
INC	International Negotiating Committee
INGO	International Non-Governmental Organisation
IPBES	Intergovernmental Science-Policy Platform on Biodiversity and Ecosystem Services
IPCC	Intergovernmental Panel on Climate Change
IPIECA	International Petroleum Industry Environmental Conservation Association
IR	International Relations
ISI	Institute of Scientific Information
JI	Joint Implementation
LULUCF	Land Use, Land-Use Change and Forestry
NASA	National Aeronautics and Space Administration
NDC	Nationally Determined Contributions
NGO	Non-Governmental Organisation
NOAA	National Oceanic Administrative Organisation
OECD	Organisation for Economic Co-operation and Development
SBSTA	Subsidiary Body for Scientific and Technical Advice
SMIC	Study of Man's Impact on the Climate
SR1.5	Special Report on 1.5°C (IPCC 2018)
STS	Science and Technology Studies
TERI	The Energy and Resources Institute
THE-QS	Times Higher Education Survey
UK	United Kingdom
UN	United Nations
UNCED	United Nations Conference on Environment and Development
UNEP	United Nations Environment Programme
UNESCO	United Nations Educational, Scientific and Cultural Organisation
UNFCCC	United Nations Framework Convention on Climate Change

UNGA	United Nations General Assembly
UNISDR	United Nations International Strategy for Disaster Reduction
USA	United States of America
USDEL	United States Delegation
USSR	Union of Soviet Socialist Republics
WCED	World Commission on Environment and Development
WCP	World Climate Programme
WMO	World Meteorological Organisation

1

Introduction

At the 24th Conference of the Parties (COP 24) to the United Nations Framework Convention on Climate Change (UNFCCC) in Katowice in December 2018, international media attention shone on the Intergovernmental Panel on Climate Change (IPCC) and its Special Report on the impacts of global warming at 1.5 °C (IPCC 2018a). The IPCC is accustomed to controversy following the publication of a report. However, this tug-of-war, over whether the latest report should be 'welcomed' or 'noted' (Allan et al. 2018: 28–29), was different. IPCC assessment reports are designed to update climate change knowledge and provide a collective basis for global negotiations at critical junctures in the UNFCCC process. This makes the organisation and the key findings of its reports objects of struggle for those wanting to delay political action. The media has often been used in these strategic attempts to undermine influential components of an assessment and its authors. The distinction between these criticisms and the struggle over the special report on 1.5 °C is that the world viewed this struggle at the site of climate negotiations and between government delegates (McGrath 2018).

As I set out to demonstrate in this book, the IPCC – as an organisation and an assessment practice – has always been shaped by the political forces of the global community's response to climate change. The IPCC established global interest in climate change and, as such, is where the politics over the meaning and collective response to the problem began. And yet, the IPCC is rarely acknowledged and studied as science situated centrally in climate politics and politics as central in and to the IPCC's formation and assessment of global climate change knowledge. IPCC scholarship increasingly documents the IPCC's role in producing objects for negotiating action or for legitimating negotiated policy decisions within the UNFCCC (Fogel 2005; Lahn and Sundqvist 2017; Livingston and Rummukainen 2020; Lahn 2021, 2022; Beek et al. 2022; Cointe and Guillemot 2023). This scholarship evidences the effect of this role on climate knowledge production, the authorship of the assessment and on the intergovernmental approval of its key findings (Petersen 2006; Hughes and Paterson 2017; Beck and Mahony 2018; Kouw and Petersen 2018; Livingston, Lövbrand and Olsson 2018; Pearce, Mahony and Raman 2018; De Pryck 2021, 2022). However, in this book, analysis begins from the IPCC as a central site in and producer of climate politics.

I came to understand the IPCC as situated centrally within and a powerful producer of climate politics through the project's central research question: who has the power to define climate change for collective response and what constitutes this power? As the organisation established and mandated to assess the latest knowledge on climate science, impacts and mitigation, the IPCC was the site to address this question. To find an answer, however, I had to look beyond the relationship between science and politics, which is central to scholarly interest in the IPCC. I too started here. However engaging with the sociology of Pierre Bourdieu (Bourdieu 1989, 1990, 1991; Wacquant 1989; Bourdieu and Wacquant 1992), forced me to situate the research question in broader global activity on climate change and study the IPCC in relation to social, political and economic interests in the problem and the struggles and forces these generate. From this starting point, the politics of climate change, the IPCC and its place negotiating a collective response look different from a model of knowledge provider for political action.

Bourdieu's notion of naming is key to how I redescribe the politics of climate change as a struggle to determine the meaning of the problem and thereby the response (Bourdieu 1986, 1991). For Bourdieu, these acts of naming are an attempt to 'fix forever' a set of power relations 'by enunciating and codifying' (Bourdieu 1986: 480). Carried within and by the name is the classificatory scheme of its origin (Bourdieu 1986). The classificatory schemes that the book interrogates are the cultural systems that determine the values and distribution of social, scientific, political and economic resources, which imprint on and in the name of climate change. Through this lens, the politics of climate change is a struggle over the social properties and material resources valued to order global relations and through which global relations are ordered. The IPCC is centrally placed within this struggle as the organisation authorised to assess the meaning and determine the practice – by whom, based on what forms of authority and through which set of activities – climate change is named. I describe the IPCC's practice of writing through the organisational actors, activities and forms of authority that have emerged over 30 years for the purpose of collectively naming this problem.

1.1 The IPCC as a Practice of Writing

The IPCC was established in 1988, with the task of assessing climate change divided between three working groups: the science (Working Group I (WGI)), impacts (Working Group (WGII)) and response measures (Working Group (WGIII)). Historical accounts of the IPCC's formation have been informed by interest in the scientific processes and politics informing the collective response (Bodansky 1993; Hecht and Tirpak 1995; Agrawala 1998a, 1998b; Skodvin 2000a). In the study of international relations (IR), the epistemic community model has been most influential. This scholarship documents the emergence of a transnational community of scientists and the conferences and workshops through which scientific understanding was transferred to a policy audience and translated into policy recommendations (Lunde 1991; Boehmer-Christiansen 1994a, 1994b; Paterson 1996; Haas 2000; Newell 2000). It is the ascendency of climate change on the political agenda, driven by the epistemic community and extreme weather events during the 1980s, that created the momentum for establishing an intergovernmental body to undertake an assessment of the state of knowledge on climate change. This established an organisation that contained both science and politics.

The epistemic community model was not designed to study an intergovernmental process that institutionalised science and politics for the production of usable knowledge (Haas 2004). For science and technology studies (STS) on the other hand, the intertwinement between science and politics in policy advice is a core focus, and the notion of boundary organisation is central to its study (Guston 2001). It is through STS concepts that much scholarly understanding and knowledge of the IPCC has been built. The notions of boundary organisation, boundary work (Gieryn 1983) and co-production (Jasanoff 2004a, 2004b) have unravelled the relationship between science and politics, documented the processes of translation between worlds and described the boundary work undertaken in bringing science and politics together and maintaining a demarcation in the organisation and its final products (Shackley and Wynne 1996; Shaw 2000, 2005; Siebenhüner 2003; Lövbrand 2007; Hoppe, Wesselink and Cairns 2013; Lidskog and Sundqvist 2015; Sundqvist et al. 2015). Re-telling the history of the emergence of climate change and establishment of the IPCC through the idiom of co-production in Chapter 2 brings to the fore the alignment between globalised knowledge and political order in how climate change became collectively known and institutionalised (Jasanoff 2004a; Miller 2004).

It is through the epistemic community literature and STS scholarship that I learned about the organisation I was studying. Holes began to appear in my grasp of this, however, during interviews. At first, it was a problem of a shared understanding, a sense that the interview respondent and I shared a framework for conceiving the IPCC, which kept the interview confined by what was known about the organisation at the time (Hulme and Mahony 2010). I revised the interview questions and asked participants to describe in detail their role in the assessment instead. This proved helpful and I began to hear how an assessment report was put together. Then a second issue arose. I learned about tasks and activities that I struggled to locate in fields of science or politics. Some interview participants had academic backgrounds in climate science and related fields, but they were not producing knowledge and assessing literature as authors or overseeing the assessment as WG co-chairs, their role was intermediary and largely administrative and technical. In fact, on a day-to-day level, they appeared to be holding the whole exercise together. This left me with the sense that I did not know what I was studying. I decided that on the most basic level, my research needed to provide a detailed account of the IPCC as an organisation.

It was during a later interview that my understanding of what the IPCC *does* was confronted. I was left feeling very uncomfortable when, for the second time, I was impatiently referred to the IPCC rules and procedure, as if the answers to all my questions were contained in that document. The problem was, despite reading this document, I could not see its significance the way my participants seemed to. That was, until I observed the rules and procedures in the making. In October 2010, I travelled to Busan in South Korea to observe the 32nd Plenary of the IPCC. It is during these annual or bi-annual meetings that the actors that I had been interviewing – delegates, bureau members and technical support unit (TSU) staff – come together for four to five days of intergovernmental decision-making. Observing this meeting was critical to understanding the IPCC as a practice of writing and the importance of social order to how climate change is written through the process.

In some respects, this meeting and the organisation I saw through it was a reflection of a particular moment in the IPCC's history. It was after the publication of the fourth assessment report (AR4) in 2007 and in the early stages of the fifth assessment cycle. The IPCC came under intense pressure in 2009, when emails between IPCC authors at the Climate Research Unit at the University of East Anglia were hacked (Pearce 2010). The email conversations between authors of the assessment were used to cast doubt on the science of climate change. Criticism further intensified in 2010, when mistakes were discovered in the AR4 on the date given for the melting and disappearance of the Himalayan glaciers (Carrington 2010). In order to address this criticism, and re-establish the IPCC's authority as the leading international assessment body, the UN Secretary-General, Ban Ki-moon, and IPCC chair, Rajendra Pachauri, requested the InterAcademy Council (IAC) 'to conduct an independent review of IPCC processes and procedures used to produce assessments' (IAC 2010a: 7). It was at the 32nd plenary that the IAC review and recommendations were discussed by the panel, and the processes and procedures for producing assessment reports were re-formulated.

On observing this meeting, I began to conceive of the IPCC as a practice of writing, both in the sense of writing climate change in and through the assessment and writing the rules by which it will be written. This meeting also enabled me to observe that not all are equal in the writing of climate change, as it became apparent that not all actors present were immersed in the proceedings or impacting its outcomes. Compared to the size of the meeting space, there was a relatively small group of countries that were actively involved in the process of revising the organisational rules and procedures. This raised questions about an actor's capacity to invest and the properties that constituted the power to shape IPCC decision-making and its products. It is through the actors, activities and forms of authority framework that I systematically explore the social properties that order relations in the organisation and through the production of an international assessment of climate change.

Through this framework, I describe the IPCC as five units: the panel, the bureau, the TSUs, the authors and the secretariat. This approach opens analysis to all actors and forms of authority, regardless of whether it is designated as scientific, political, technical or administrative, as all of these activities are required to put together a global assessment of climate change. This approach makes it possible to explore the relationship between these activities and participation in the IPCC and the economic investment that becoming a symbolically powerful writer of climate change is dependent upon. Bourdieu's notion of capital is critical to this (Bourdieu 1986). To identify the properties distinguishing actors within the IPCC and to explore the relationship between this social order and the global distribution of resources, I retained Bourdieu's concept of capital.

Capital makes it possible to identify and unpack what constitutes authority within the panel, bureau, TSUs, secretariat and authorship of the assessment – the distribution of social, scientific, political and economic resources that govern an actor's access to, location within and influence over the organisation and its assessment practice. Although Bourdieu identified three principal types of capital – economic, cultural and social (Bourdieu 1986; Wacquant 1989, 1998) – the valued properties and their capacity to order relations in the IPCC had to be identified empirically through participant observation and interviews. To understand the symbolic power these forms of authority have in the IPCC's writing of climate change in the present, it was necessary to return to the historical emergence of the organisation. It was during the establishment of the IPCC in 1988 that the cultural foundations were laid, which in turn identified and distinguished the properties that would be organisationally valued and the actors that embodied these. I use the actors, activities and forms of authority framework to describe the social order at each stage of the assessment's production, from member government's decision to repeat the process (Chapter 5), the scientific assessment (Chapter 6) to intergovernmental approval of the report's key findings (Chapter 7). It is by following the assessment report and mapping the social order of its conduct that the book addresses its central question and explores how the global order of relations imprints on and through the naming of climate change.

1.2 The Method of Data Collection

The conceptualisation of the IPCC as a practice of writing and the development of the actors, activities, forms of authority framework developed through in-depth empirical study over 15 years. There were several layers to my immersion in the organisation. I began with the reports themselves, reading the Summary for Policymakers (SPM), the chapter executive summaries and recording the names, affiliation and nationality of the authors for the first (1990), second (1995), third (2001) and fourth assessment reports (2007). This gave me a sense of disciplinary constructions of climate change, and I began to recognise the names of key actors that served on multiple assessments as authors and bureau members. I contacted these actors for interview and began interviewing in the summer of 2009. My approach to interviews changed rapidly in the beginning, when I was learning from each conversation and at the same time, struggling with the sense that I understood less about the organisation with every interview. It was the concern that my interviews were providing more data on what people thought about the organisation than what the organisation is and does that led me to attempt to immerse myself in the undertaking of an assessment as my respondents were. I began to ask interview participants to describe what they did, step by step, in the assessment, and from this, I began to build up a detailed picture of how an IPCC assessment report is made.

I have undertaken over 40 interviews in total and had many more conversations and email exchanges to check and refine the details of the IPCC's assessment practice. However, I could not have described the social order shaping these activities and their imprint on the final product without observation. I was increasingly hearing about the importance of the TSU in the assessment's production, and I expanded my field research to conduct further interviews and visit the TSU for WGII's contribution to the fifth assessment report (AR5) at the Carnegie Institution for Science at Stanford University. Later, in 2019, I also visited the WGIII TSU at Imperial College London during the sixth assessment cycle (AR6). During observation of the 32nd plenary of the IPCC in October 2010, I began to quantify the asymmetry in participation by logging and timing each intervention (see Table 4.2). After the meeting, I continued interviews, which became increasingly focused on the finer details of putting together an assessment and about the asymmetries observed during the plenary.

Since the initial PhD study of the IPCC, I have expanded data collection through collaborative research projects that have helped to provide further quantification of asymmetries. This includes a social network analysis (SNA) of institutional affiliation and co-authoring patterns of WGIII authors in the AR5 (Corbera et al. 2016). One of the gaps from the initial interview data was that it was the view from developed country participants (see Appendix 1). To address this in a subsequent SNA-informed study, we designed and conducted a survey of AR5 WGIII authors to develop a more intricate understanding of the forms of authority ordering author relations in the assessment (Hughes and Paterson 2017). It became increasingly important to situate this data in the broader global knowledge landscape of the climate field. For this, I began to attend UNFCCC COP meetings, including COP 23 and COP 24, where I observed the formation of the Local Communities and Indigenous Peoples Platform. Then in January 2022, a successful grant application, initiated a more detailed study of the AR6.[1] This has included observation of WGII and WGIII's virtual approval sessions and observation at COP 27 in November 2022 and a Subsidiary Body meeting (SB58) in June 2023 to follow the dissemination of the AR6 and its uptake in the Global Stocktake.

The account provided of the IPCC's practice of writing climate change, the social order this is built upon and its imprint on the naming of climate change is informed by all forms of data collected through each stage of research. The book's detailed description of the organisation and its practice for putting together an assessment report is informed by and in reference to interviews, IPCC documentation, Earth Negotiation Bulletin (ENB) reports of meetings and the scholarly literature. In order to demonstrate quantitatively the asymmetries in participation, I use IPCC participant lists, author lists, government review comments and the ENB reports, as well as the IPCC's own studies and analysis of developing country participation. This enables me to provide a detailed and quantitatively supported study of who participates in IPCC meetings, what enables meaningful participation and with what effect for how we know and respond to climate change.

1.3 Contribution to Knowledge

By deconstructing the IPCC through the actors, activities and forms of authority framework and describing the practice of writing climate change I offer a novel way to understand this organisation and its place in climate politics. The book provides a detailed account of the historical emergence of the IPCC's practice of writing and the cultural properties that were valued to order relations, how this order imprints on IPCC products and how this order is challenged and changes over and through each assessment cycle. Through this account, I make a contribution to existing literature on the IPCC and the study of intergovernmental

[1] ESRC grant application on the Politics of Science in Climate Cooperation led by Patrick Bayer (University of Glasgow) and in collaboration with Erlend Hermansen (CICERO, Norway) (ES/W001373/1).

organisations more broadly. On a practical level, the book contributes to understanding the relationship between measures of authority and global resource distribution and its impact on developing country participation. The practical utility of this approach is that it identifies actors and activities by which social order can be challenged and developing country participation strengthened. This remains critically relevant to the IPCC and its place in the collective stocktake of the Paris Agreement, and to ensure that all new global knowledge bodies design for participation by all from the outset.

Understanding the IPCC as a practice of writing and disaggregating this practice into the actors, activities and forms of authority constituting it is inspired and underpinned by the sociology of Pierre Bourdieu. Bourdieu's scholarship has emerged as influential in the study of IR (Jackson 2009; Bigo and Madsen 2011; Leander 2011; Adler-Nissen 2013). It has left a particular mark in elevating the analytical significance of practice (Neumann 2002; Pouliot 2010, 2016; Bigo 2011; Drieschova and Buerger 2022) and in illuminating the culturally constituted symbolic forms that power takes (Williams 2007; Adler-Nissen 2013; Eagleton-Pierce 2013; Adler-Nissen and Pouliot 2014; Hughes 2015, 2023). From Bourdieu I learned how to study the making of social order (Bourdieu 1988, 1989, 1990, 1998). His analytical tools provided a unique way to isolate the IPCC as an organisation for internal study, while ensuring analysis is situated in broader social and political struggle over climate change. In identifying the actors that have the power to determine the meaning of climate change – the symbolic power – to name, and what constitutes this power, the book builds on existing Bourdieu-informed scholarship on authority (Sending 2015), symbolic power (Eagleton-Pierce 2013) and pecking orders (Pouliot 2016), while contributing to further systematic analysis of the properties of this power through the notion of capital.

In identifying the IPCC as a practice of writing, the book aspires to do what other authors in the study of IR and beyond have done through observation, ethnography and narrativisation of practice, that is, to show how organisational practices contribute to making the object of their activity (Riles 2000; Hull 2012; Raffles 2014; Sending 2015). As Hugh Raffles demonstrates through his account of the Amazon, the Amazon became an object of British life (and beyond) through the imperial scientific practices that aimed to establish it as a site of discovery and set out to record and claim it as such; practices that have and continue to shape both the nature and knowledge of the Amazon in tangible and lasting ways (Raffles 2002). Putting this into the context of international organisational life, Riles' study brings to light the role that information plays in bringing objects like 'the environment' and 'women' into existence, which also becomes the means by which they exist (Riles 2000: 179). Sending's study on peace building, on the other hand, illuminates how the content of these governance objects is also shaped by actor attempts to constitute and struggle over the authority to control and govern them (Sending 2015). The contribution I hope to make to these studies of the duality of knowing and acting on through recording, gathering, networking and documenting is the systematisation of the study of power relations. This book aims to demonstrate concretely how these activities are marked by social order and how social order leaves its mark through these naming practices.

The analytical framework of the book makes it possible to document the extent of the involvement of member government in the IPCC's practice of writing climate change. The intergovernmental nature of the IPCC as a knowledge provider has been of interest to scholarship from the outset. The establishment of an intergovernmental body over a science-governed assessment process has been viewed as an attempt to gain control over climate science as it ascended the political agenda (Haas 2004). For actors within the IPCC and those observing it closely, the intergovernmental character and the involvement of member governments in the organisation and assessment practice, particularly during the approval of the SPM, guarantee the utility and impact of the knowledge base on climate negotiations (De Pryck 2022). However, the extent of member government involvement and the distribution of power in and over the practice for producing the assessment and its final outcomes have not been documented until now. This account describes the activities of member governments as focal points and delegates at each stage in the production of the assessment, which is particularly revealing at the start of the assessment when governments inform the direction and content of the next global report on climate change. As documented in Chapter 5, starting with the decision to repeat the process through the election of the bureau, scoping the report, commenting on drafts and approving the final outline, member governments are deeply invested in attempting to shape the content and limit the implications of the next report.

Describing the extent of member government activities required for meaningful and impactful participation in the IPCC's practice of writing begins to discern the dependent relationship between economic resources and symbolic power to write climate change. Participation in the IPCC is a resource-intensive process that requires significant economic and human resource investment over time. It is through fulfilling all the necessary activities and tasks as government delegates to the panel and national focal points to the organisation that the IPCC's practice of writing is learned, avenues for influence open and symbolic power is gained. This understanding provides the basis of exploring the continuing asymmetries between developed and developing country participation despite organisational efforts over 30 years to deepen engagement. Combined, the description of member government activities and the mapping of power relations make the IPCC an ideal teaching site for unpacking the relationship between science and politics and simulating the intergovernmental negotiation of the SPM. Alongside the book, I offer an outline for role playing the politics of approval to enable instructors to develop a lesson plan relevant to their particular setting and learning requirements. This is available on www.cambridge.org/hughes.

At the same time, the analytical framework of the book helps the reader to look beyond the focus and relationship between science and politics in present IPCC scholarship. In this approach, all actors, activities and forms of authority are subject to empirical study to determine their role in and influence over the IPCC's practice of writing. This identifies the TSUs that sit alongside the WG co-chair as critical in the realisation of the assessment. Through the everyday activities of scheduling, emailing, compiling, formatting and editing, actors within the TSU acquire the most intimate knowledge of the assessment and its progression. This is a valuable form of cultural capital, which translates into symbolic power in and over the writing of climate change, and the book explores who has access to these resources and how they are distributed within the panel.

This analytical approach has utility for studies of international organisations where generating knowledge and expertise is central to the organisational mandate. The power of the bureaucracy in the study of international organisations has been brought to the fore in the study of IR, challenging disciplinary conceptions of what actors and forms of authority matter (Barnett and Finnemore 1999). Within the study of global environmental politics, the influence of treaty secretariats has become an important area of study, and detailed comparison between these has been undertaken (Yamin and Depledge 2004; Bauer 2006; Depledge 2007; Biermann and Siebenhüner 2009; Jinnah 2014). This framework could significantly contribute to further opening up the black-boxed nature of the secretariat, as well as revealing what other units within the organisation undertake and compete for administrative tasks. Combined with the importance of learning the overriding practice and purpose of an organisation – as in the practice of writing – this approach could be particularly insightful in studying how the social order of an organisation is imprinted and re-made through the forms of expertise and organisational products it generates (Adler and Bernstein 2005).

The contents of the book have implications for the design of new knowledge bodies to inform treaty-making. The IPCC model has already proven influential in the design of other global assessment bodies, including the Intergovernmental Science-Policy Platform for Biodiversity and Ecosystem Services (IPBES) (Larigauderie and Mooney 2010). Negotiations are underway for the formation of a science-policy body on chemicals and waste (Wang et al. 2021), and the same is likely to arise in the newly negotiated agreement on the conservation and sustainable use of marine biological diversity in areas beyond national jurisdiction (BBNJ) (Tessnow-von Wysocki and Vadrot 2020). If from the outset, the architects acknowledge that this body is and will be a site of politics and make meaningful participation by all a central objective, a body can and will be designed differently. I discuss the implications of the findings of the book for new science bodies in concluding the book in Chapter 8.

The book's account also reveals its limits and where future study can build on and develop understanding in advancing participation by all. This is most notable in relation to the struggle over the categorisation of developed and developing countries that begins with the approval of the outline in Chapter 5 and is documented through the politics of approval in Chapter 7. Through recounting these events, it becomes apparent that I am describing an order of relations that is not as it was when the IPCC was established in 1988. As the practice of writing reflects changing global order, so too must critical scholarship find the analytical means to record and illuminate these shifts. In the book, I use the IPCC's own categorisation of developed and developing countries to study participation. This proves effective in revealing the continued dominance of developed countries in the global knowledge economy and, by extension, the power of scientific culture to order relations in the authorship of the assessment. However, it also increasingly masks asymmetries within the developing county category, which is why some countries fought so hard to erase it from the assessment (Chapter 7). Future studies of the IPCC and climate politics more broadly need to find a way to carefully unravel these tensions and more productively understand and engage with the changing dynamics of participation in all agreement-making processes.

Relatedly, there is also a need to more strongly relate the order of relations in the practice of writing climate change to greenhouse gas (GHG) emissions, revenue from fossil fuels and dependence on fossil fuels. Sometimes in the struggle over scientific and technical details and the strategies deployed, the interests and interventions of the underlying drivers of member governments can become obscured. Research can help to make the apparently complex struggles simpler and contribute to rendering clearer what exactly countries are negotiating for (with what resource base) in climate agreement making and at what cost to all peoples and the planet.

1.4 The Journey of the Book

In this book, I hope to take you, the reader, on an intricate journey into the practice of writing climate change, the social order through which it is written and the imprint this leaves on how we know and act upon climate change. I begin this journey in Chapter 2, identifying some of the most important conceptual resources available in the study of environmental problems to address the central question of the book, namely who has the power to write climate change and what constitutes this power. This allows us to explore the early history of the IPCC from different approaches and models of the relationship between science and politics. By Chapter 3, however, it is apparent that the history of the IPCC's emergence can be re-told again through a Bourdieu-inspired approach, whereby climate politics is a struggle to determine the meaning of the problem and the order of relations preserved or challenged through the name approved. This situates the IPCC centrally within global climate activity as the organisation authorised to write the meaning of climate change and the rules by which it can be written.

With the IPCC situated in the global field of climate activity and the forces this generates, I turn inward in Chapter 4 to the actors, activities and forms of authority that constitute this organisation and its practice of writing climate change. History is central to documenting the emergence of the social order within the organisation, and the chapter provides an account of how the IPCC has arrived at its present form. With a more intricate understanding of the organisation and the order of relations shaping it, the book begins its journey along the assessment production pathway from the decision to repeat the process (Chapter 5), through the scientific assessment (Chapter 6) to the politics of approving the report's key findings (Chapter 7). Following me on this intricate journey makes apparent the stakes for all in the IPCC's practice of writing climate change and the power of this order-making problem to make and shape politics in its name.

2

Knowledge, Power and Order in the Construction of Environmental Politics

In 1985, the Villach Conference produced a statement that acknowledged that 'While some warming of climate now appears inevitable due to past actions, the rate and degree of future warming could be profoundly affected by governmental policies on energy conservation, use of fossil fuels, and the emission of some greenhouse gases' (WMO 1986). This conference is identified as a pivotal moment in the emergence of climate change on the political agenda. The statement raises questions about shared constructions of environmental degradation and questions of power: who wrote this statement, on the basis of what authority, and how was it acted on? What was the role of scientists and scientific knowledge in identifying this problem? What other actors and social, political and economic forces structured this construction of the issue?

These were the questions I started out with, and the Intergovernmental Panel on Climate Change (IPCC) was the site to study them. For this, I needed a method – conceptual and methodological tools – that would enable me to explore how and by whom climate change is constructed for social and political action. I started with scholars that shared an interest in the problematisation of environmental degradation, examining the frameworks they developed to undertake empirical research and inform their analysis. This chapter charts my journey through the key scholars and approaches that provided the foundations for this book. The review takes us through some of the most influential and comprehensive frameworks for identifying the role of science and scientists in environmental degradation in international relations (IR) and science and technology studies (STS), which include the epistemic community model, discursive and ideational frameworks and the idiom of co-production.

Scientific knowledge identifies the causes and consequences of shared environmental issues and as such, empowers new sets of actors and forms of authority in international politics. The epistemic community model provides a framework for delineating the role that transnational communities of scientists play in defining an issue for political response (Haas 1989, 1990) and provides the basis for scholarly accounts of the emergence of climate change on the political agenda and the institutionalisation of climate science within the IPCC (Lunde 1991; Boehmer-Christiansen 1994a, 1994b; Paterson 1996; Newell 2000). Revisiting these early histories initiates the book's journey into the IPCC and enables me to retell the emergence of climate change as an object of politics. However, criticism of the epistemic community model has been equally important in furthering understanding of the relationship between knowledge and politics in the construction of global environmental action, particularly for bringing power and who holds it into focus. The work of Karen Litfin (1994) on ozone discourses and Maarten Hajer (1995) on the framing of acid rain shifts analytical attention from expert communities to discursive practices. These approaches introduce new sets of actors and highlight the degree of contestation and struggle in social construction processes. While these approaches do not add to knowledge of climate politics or the IPCC per se, they set important standards for putting together the book's analytical framework. To bring climate change back into view and to explore the power of economic and social order in its construction, the chapter turns to the normative framework developed by Steven Bernstein (2001) and the STS idiom of co-production and its application by Clarke Miller (Jasanoff 2004a; Miller 2004). These approaches bring broader patterns of social organisation and their effect on the IPCC and its assessments of climate change into focus. Thus, by the end of the chapter we have a more complex cast of actors and processes involved in the problematisation of climate change than from where we started, but questions around the properties of power and their distribution remain.

2.1 Power versus Science: The Epistemic Community Model

The epistemic community model arose from interest in how and why cooperation between states occurs. Peter Haas sought to understand why, despite differences over who should pay and reluctance to forsake short-term economic welfare, states cooperated in environmental regimes where there were no clear mutual interests between states or guarantees that protection costs would be equally distributed (Haas 1990: 347). Haas suggested that international environmental regimes stemmed as much from transnational communities of shared knowledge, or epistemic communities, as state power (Haas 1989, 1990, 1992a, 1992b). An epistemic community is defined as 'a network of individuals or groups with an authoritative claim to policy-relevant knowledge in their domain of expertise' (Adler 1992: 101). Members of these communities are said to: (1) share knowledge about the causation of social and physical phenomena in an area for which they have a reputation for competence, (2) have a common set of normative beliefs about what will benefit human welfare in such a domain, and (3) share a common policy project. The expertise of an epistemic community is in particular demand in instances where there is uncertainty about the physical characteristics of an issue and how best to politically pursue a response. In these cases, the epistemic community helps states to identify their interests, frame the issue for collective debate, propose specific polices and identify salient points for negotiation (Haas 1992a: 2). In this approach, power – although not designated as such – lies in the community's authority to define the problem for political response, which may be further consolidated by the institutionalisation of the scientific view within national administrations and international secretariats (Haas 1992a: 4).

The epistemic community model has proven influential in interpreting the emergence of climate change as an international political issue (Lunde 1991; Paterson 1996; Haas 2000; Newell 2000). Matthew Paterson (1996: 144) concludes that 'the international development of climate as a political issue ... can plausibly be interpreted in terms of the effect of the development of an epistemic community on the subject'. Further claiming that, '[I]n the IPCC we can see the epistemic community at its most organised' (Paterson 1996: 146). Even those more critical of this approach acknowledge the role of an epistemic community in raising the political profile of climate change (Vogler 1995: 204; Bernstein 2001: 161). Revisiting these accounts provides useful historical background on the emergence of the IPCC and enables an exploration of how science and scientific knowledge have been theorised in the positioning of climate change on the political agenda.

Scientific interest in the effect of atmospheric gases on the global climate has a long history, dating back at least as far as 1824, when the French philosopher Jean-Baptiste Fourier hypothesised that the atmosphere trapped heat in a manner similar to a 'hothouse', or greenhouse (Weart 2008). The heat-trapping capacity of atmospheric gases, such as carbon dioxide (CO_2) and water vapour, was first tested by the female scientist Eunice Foote, who presented her findings at an annual meeting of the American Association for the Advancement of Science (AAAS) in 1856 (Sorenson 2011). Credit is usually given to John Tyndall for this initial discovery, who observed the same effect through laboratory experimentation in 1859 (Weart 2008). The role of heat-trapping gases in the atmosphere and their effect on the prevailing temperature was elaborated over the nineteenth century, and by 1886 the Swedish scientist Svante Arrhenius had quantified the contribution of carbon dioxide (CO_2) to the heat balance of the earth, indicating that a doubling of atmospheric CO_2 concentrations could increase average global surface temperature by 5.7°C (Bolin 2007: 4). Despite these discoveries and a series of papers on fossil fuel emissions and climate change by G. S. Callendar in the 1940s and 1950s, it took advances in computer modelling, rising CO_2

concentrations and increased support for international scientific cooperation to initiate sustained scientific interest in human's impact on the climate (Paterson 1996; Shackley and Wynne 1995; Haas 2000; Edwards 2001, 2010; Miller 2001a; Weart 2008; Zillman 2008).

One of the first scientific assessments of 'the possible impacts of man's activities' on the climate was generated by a three-week international scientific gathering held in Sweden in 1971 (SMIC 1971). This report is said to have become 'required reading' for participants at the UN Conference on the Human Environment held in Stockholm the following year (Kellogg 1987: 121). This conference represented mounting concern over human's negative impact on the environment, including human's potential to alter the climate. It called for increased research and monitoring of CO_2 build-up and established a new UN agency for the preservation of the environment: the United Nations Environment Program (UNEP) (Agrawala 1999: 159). In 1979, the first World Climate Conference was held in Geneva, organised primarily by World Meteorological Society (WMO), it provided a major international platform for scientific interest in climate change (Agrawala 1998a: 607). The conference recognised climate change as a serious problem and issued a declaration calling on the world's governments 'to foresee and prevent potential man-made changes in climate that might be adverse to the well-being of humanity' (WMO 1979). Efforts were also initiated to create an international climate research programme, which eventually led to the creation of the World Climate Programme (WCP). Co-sponsored by the WMO, the newly formed UNEP, and the International Council of Scientific Unions (ICSU), the WCP was the first internationally coordinated program of climate research and proved critical for fostering greater scientific interest, furthering research into climate change and supporting the translation of scientific knowledge into political concern (Paterson 1996: 28–29).

International scientific conferences and workshops continued into the 1980s. As the focus of these events shifted towards the social and political implications of climate change the events increasingly sought and attracted the attention of a diversified range of actors, including the policy oriented. Although different explanations are cited, authors agree that the 1985 Villach Conference was pivotal (Bruce 1991; Boehmer-Christiansen 1994a; Hecht and Tirpak 1995; Jaeger and O'Riordan 1996; Franz 1997; Haas 2000; Bernstein 2001). Co-sponsored again by UNEP, ICSU and the WMO, the Villach Conference is said to represent the core of an epistemic community (Bernstein 2001: 162) and to have initiated the politicisation of climate change (Paterson 1996: 29). At this conference the 89 participants from 29 developed and developing countries and three sponsoring organisations confirmed global warming trends (WMO 1986), and there was an apparent 'shift of emphasis' away from more research required towards assertions of the need for political action (Paterson 1996: 31). The report that emerged from this conference was widely disseminated and formed the basis of the Brundtland Commission's recommendations in *Our Common Future* on action to protect the earth's climate (Franz 1997: 22; WCED 1987).

The 1985 Villach conference also initiated the establishment of an international scientific committee, institutionalising some of the most proactive members of the epistemic community. UNEP's director, Mostafa Tolba, first tabled the idea of an advisory panel to guide climate change policy at the Villach conference (Agrawala 1999: 160). The committee was to explore policy options for responding to climate change, set research priorities and conduct assessments of the long-term impacts of climate change (Agrawala 1999). Tolba's idea was embraced by several conference participants, and in 1986 the Advisory Group on Greenhouse Gases (AGGG) was established under the auspices of WMO, UNEP and the ICSU (Agrawala 1999: 160–61). The AGGG was composed of a group of seven experts that were each involved in the parent bodies, had long-standing scientific careers and were linked to national bureaucracies (Agrawala 1999). It was envisioned that the AGGG would design and implement 'constructive interventions into energy, climate, and socio-economic areas' (WMO 1986: 43 in Agrawala 1999: 161). Although this first attempt to institutionalise scientific advice on climate change would be overshadowed by the IPCC (Agrawala 1999), the AGGG performed its role according to the epistemic community model – devising and disseminating climate change policy response options through support of workshops and conferences.

Two such workshops were arranged for 1987, the first of which is said to have advanced scientific understanding of the regional impacts of climate change, and the second, the policy discussions on mitigating these effects (Franz 1997). As the focus of the community's efforts shifted towards the policy implications of climate change the professional backgrounds of the invited participants began to diversify, with an increasing number of policy-oriented actors in attendance. At the first workshop in Villach, 48 participants from academia, environmental advocacy groups and some national environmental agencies attended. The second workshop, held in Bellagio, Italy, was policy focused and brought together new policy-oriented actors, including representatives from UNEP, environment departments in the United States, the Netherlands and the European Union, and NGO actors (Franz 1997: 23). The workshop indicated that policies were necessary to keep temperature and sea level increases within 'tolerable rates', with 0.1°C/decade suggested as a suitable global threshold for temperature increase (Franz 1997: 23–24). The group also calculated emission reduction targets, with a 66% reduction in CO_2 suggested and proposals on how to achieve this (Franz 1997).

The largest and most influential policy audience on climate change gathered in 1988, at the Toronto conference on 'The Changing Atmosphere: Implications for Global Security'. Three hundred and forty-one delegates attended the conference, representing 46 countries and 24 international organisations (Franz 1997: 25). Amongst the most notable of the policy audience were Gro Harlem Brundtland, the Norwegian prime minister and leader of the Brundtland Commission on Environment and Development (WCED), the Canadian prime minister, and a number of G7 ministers (Agrawala 1999: 162–63). Participants were provided with a common framing of climate change through a background paper informed by the Villach and Bellagio workshops and written by a member of the AGGG. The final conference declaration called for a 20% reduction in OECD (Organisation for Economic Co-operation and Development) emissions from 1988 levels by 2005, making it the most significant climate policy initiative at the time (Agrawala 1999: 169).

It was not only scientific initiatives that were significant in raising the profile of climate change. In 1988, a number of other events helped to move climate change to a topic of broader social and political concern. Two of those events occurred in June. The first was a statement made by NASA scientist, James Hansen, in US Congressional hearings that were convened due to unusually hot and dry summer conditions (Hecht and Tirpak 1995: 383–84; Jaeger and O'Riordan 1996: 16). Hansen stated that he was 99 percent certain that the warming of the 1980s was not a chance event. Unexpected patterns of drought, floods and other extreme weather occurrences were also being experienced globally, including in the USSR, Africa, India, China, Brazil and Bangladesh (Boyle and Ardill 1989: 1–4). These events increased political interest and media attention in the scientific predictions of global warming and in the Toronto conference, which received high levels of media attention as a result (Franz 1997: 25–26). This momentum was built on by Malta when they raised climate change as a matter at the UN General Assembly. By December 1988, a resolution had passed (UNGA RES/43/53) endorsing the establishment of the IPCC.

There are disparities between scholar's accounts of the establishment of the IPCC. Some credit UNEP's Mostafa Tolba and his letter to the United States Secretary of State with the IPCC's formation (Hecht and Tirpak 1995; Agrawala 1998a; 1998b). Others highlight that UNEP was more interested in a framework convention than a scientific panel, suggesting instead that the IPCC emerged from debate and corridor consultation at the 1987 World Meteorological Congress in Geneva (Zillman 2007: 870–71, 2008: 27–28). Officially, it was after consultations within and between the WMO congress and the UNEP governing council that a co-sponsored intergovernmental assessment panel on climate change was agreed (Bolin 2007: 47). The critical feature of this newly established body was its intergovernmental nature, which meant that the organisation created for generating international assessments of climate change was a governmental and scientific process. Divisions between departments appear to have contributed to an intergovernmental process over the science-led organisation of previous assessments. Although the Environmental Protection Agency and Department of State were supportive of a convention process, the Department of Energy (DoE) opposed policy action and was critical of the Villach outputs because government officials had not been involved (Hecht and Tirpak 1995: 380–81). The outcome of discussions between these parties was a US proposal for 'an intergovernmental mechanism' to conduct a government-led, scientific assessment of the climate change issue (Agrawala 1999: 611).

Thirty countries accepted the WMO Secretary General's invitation to the first session of the IPCC, including 11 developing countries (IPCC 1988). This session was concerned with formalising the structure and function of the panel, although many of these decisions had been prepared prior to the session (Bolin 2007: 49–50). The work of the panel was divided into three main areas: science, impacts and response strategies. The tasks of the three working groups were elaborated during the session and the IPCC chair and WG chairs were elected (Zillman 2007: 873). This process institutionalised key members of the epistemic community responsible for raising the political profile of climate change in the 1980s, most notably the newly elected IPCC chair Bert Bolin, and supported the claim that this model explains the source of new international institutions (Adler and Haas 1992).

Although the epistemic community model has been used to explain and explore the origins of the IPCC and the politicisation of climate change, Haas (1990) is sceptical of its applicability to this issue area. He identifies a number of factors inhibiting collective action on climate change driven by epistemic consensus, including the recalcitrance of the United States, the cost of action and the unequal distribution of costs between states (Haas 1990: 358–59). Haas has been particularly sceptical of the IPCC's role in fostering epistemic consensus, claiming that the intergovernmental nature of the IPCC stifles the epistemic community's ability to function as theorised. In fact, Haas considers the IPCC an attempt by governments to gain control over the scientists and the diplomatic process, which had ascended too quickly up the political agenda in the 1980s under the epistemic community's influence (Haas 2000, 2004; Haas and McCabe 2001). He suggests that the intergovernmental design of the panel makes it difficult to operate and implement independent initiatives to progress collective climate action. For instance, the IPCC chairman is elected by member governments, which gives the panel the power to prevent the appointment of effective epistemic community members (Haas 2004: 581).

The failure of the epistemic community to engender greater international political action on climate change indicates the complex social and political components of understanding, defining and treating the climate change issue, complexities recognised as making it a much harder case for multilateral diplomacy than other environmental issues (Haas 2008: 2). By the mid-1980s the climate change community had framed climate change for collective debate, proposed specific policies and identified salient points for negotiation (SCOPE 1986; WMO 1986). It is not that the scientific community failed to provide a definition of climate change, rather a complex interplay between scientific, economic and political dynamics in the conceptualisation of climate change resulted in the rejection of the scientist's definition of the problem and proposed policy options. However, this account of the politicisation of climate change and the IPCC's establishment also indicates weaknesses in the underlying assumptions of the epistemic community model, particularly concerning the role of science in the construction of political issues. Confronting Haas's account of truth and power reveals some of the limitations of focusing on epistemic communities in analysis of the problematisation process.

The misgivings Haas documents in relation to the IPCC are informed by his view that knowledge can improve politics, a theme that is developed throughout his work on social learning and its agents: epistemic communities (Haas 2000, 2004; Haas and McCabe 2001). Haas is concerned with 'usable knowledge': scientific knowledge that is accurate and politically tractable to politicians and policymakers (Haas 2004: 572). Authority and legitimacy are vital constituents of usable knowledge, and in order for it to be recognised as such, Haas stresses that the institutional processes for developing usable knowledge must remain insulated from political interference (Haas and Stevens 2011). According to this account, the knowledge generated by the IPCC should have taught the decision-makers that cooperation on an international agreement to reduce greenhouse gases was in their best interests. Why, then, after 30 years of the IPCC, six rounds of assessment reports, and a Framework Convention, are emissions still rising? For Haas, the answer lies in the intergovernmental nature of the IPCC, which failed to separate truth from power and, as such, has not produced legitimate, usable knowledge. Haas concludes that the IPCC has been designed 'to keep science on a tight leash and, not surprisingly, IPCC scientists have been unable to exercise sufficient discretion to develop more politically tractable advice' (Haas 2005: 396).

Haas's view of knowledge and politics has been strongly criticised for its rationalistic assumptions. These criticisms reveal a number of limitations in how epistemic communities are theorised to function. Firstly, the epistemic community model assumes that it is both possible and preferable to separate scientific knowledge from social and political processes (Lidskog and Sundkvist 2015). Secondly, the model assumes the knowledge transfer pathway and social learning are unidirectional: science educates politics (Newell 2000). Thirdly, the model does not consider the interests of the epistemic community or the competition between communities of experts and how these dynamics shape constructions of the problem (Bernstein 2001). Taken together, the assumption is that scientists and scientific knowledge diagnose environmental problems in the absence of social, political and economic forces. However, scholars have demonstrated how politics impinges on scientists in the form of self-censorship and knowledge selection in compiling IPCC assessment reports (Newell 2000). This leads Newell to conclude that the knowledge/power transfer from scientific expert to policy community runs both ways, rather than the linear one-way transfer implied by the epistemic community model (Newell 2000: 42).

The work of Litfin (1994), Boehmer-Christiansen (1994a, 1994b), Newell (2000) and Bernstein (2000, 2001) highlights the complex relationship between social, political and economic forces in the conceptualisation of environmental issues. Bernstein (2001: 174), for example, highlights that scientists were not the only authoritative experts interested in climate change and development, as environmental economists also sought, or were solicited, to influence policy formulation. The epistemic community model both overlooks the interests of the scientific community and assumes the professional background of these actors to be analogous prior to study. As Wendy Franz's (1997) account of the Villach conference makes apparent, in many instances scientists were unwilling to translate the science of climate change into policy response options for fear of appearing too close to policy and thereby undermining their scientific authority. Instead, over the course of the 1980s, the community interested in climate change expanded and diversified, with policy-oriented actors attending workshops and conferences alongside the climate science community.

Finally, when the epistemic community is the central unit of analysis, the workshops, conferences and assessments that bring these actors together in practice are overlooked. And yet, it is actor's participation in these activities that constitutes them as epistemic community members and legitimate them as recognised international climate experts, making these sites critical to the community's formation and to the formation of a shared understanding of climate change. These activities are not simply a component of the history of climate science and politics – conferences, workshops and assessments are constitutive of how climate change has become known and acted upon collectively. In the following section, I look to alternative approaches for theorising knowledge and power in the construction of environmental problems that account for a more entangled relationship between scientific knowledge and political response.

2.2 The Power of Discourse

Seeking to explain by whom and through what processes ozone depletion was framed for and by the treaty process, Karen Litfin (1994) was the first to adopt a Foucault-inspired discourse analytical approach to study the social construction of global environmental issues. Litfin understands discourse as a set of linguistic practices and rhetorical strategies embedded in a network of social relations (Litfin 1994: 3). She uses this definition of discourse to interrogate the role and power of science and scientists in the treaty formation process and observes that while scientists played a facilitative role in initially framing ozone depletion, many were reluctant to step into the policy arena by making normative judgements about the social implications of their research. She also observed that once the scientists had produced this knowledge, it 'becomes something of a collective good, available to all who want to incorporate it into their discursive strategies' (Litfin 1994: 37).

According to this approach, power does not necessarily lie with those who produce knowledge; it also has the potential to empower those who make use of it – interpreting and manipulating it in their frame of the problem. Litfin conceptualises those responsible for shaping the discourse on ozone depletion as *knowledge brokers*: intermediaries between the scientists who produce the knowledge and the policymakers who consume that knowledge (Litfin 1994: 4). This is an important addition to the epistemic community model and runs counter to its causal logic – it is knowledge brokers that translate scientific knowledge, which Litfin suggests underscores that interpretation is more important than scientific fact (Litfin 1994: 37). Litfin concludes that the dominant discourse that emerged around ozone depletion was a powerful determinant of what could and could not be thought, delimiting the range of policy options and serving as 'precursors to policy outcomes' (Litfin 1994: 37). Her empirical investigation also indicates that while scientific knowledge may facilitate cooperation, the production and interpretation of knowledge is a political process. Thus, far from rationalising politics, knowledge of ozone depletion fed into new and existing arenas of political contestation (Litfin 1994: 19).

The discursive approach and analytical focus of Maarten Hajer's (1995) work, on the other hand, shifts attention from the role of scientists and scientific knowledge in treaty making to society's problematisation of environmental degradation more broadly. Hajer seeks to understand how issues are defined in the policy process and how emergent environmental discourses constrain political action and enable social change. He develops a Foucault-informed discursive analytical framework, conceiving of politics as a struggle for discursive hegemony, whereby actors try to convince others of their definition of reality (Hajer 1995: 59). Hajer understands discourse as 'a specific ensemble of ideas, concepts, and categorizations that are produced, reproduced, and transformed in a particular set of practices and through which meaning is given to social and physical reality' (Hajer 1995: 44). In this approach, the institutional context is as important as language because it 'codetermines what can be said meaningfully' (Hajer 1995: 2). He identifies environmental discourse as fragmented and contradictory: 'an astonishing collection of claims and concerns brought together by a great variety of actors' (Hajer 1995: 1). Despite this, however, coherent definitions of the causes and consequences of environmental problems emerge, and Hajer seeks to illuminate how discursive order is created and maintained.

Hajer's study describes how the conceptualisation of environmental problems changed over time. When environmental degradation first emerged on the international political agenda in the 1970s, pollution was not considered a structural problem but rather as something that could be controlled and contained through quality control targets and abatement strategies (Hajer 1995: 25). Hajer identifies the environmental discourses that came to challenge this way of thinking, from those that popularised the notion of ecological crises, such as the *Limits to Growth* thesis (Meadows et al. 1972), to those that held modern society responsible for systemic environmental degradation, as captured by the text, *Small is Beautiful* (Schumacher 1974). While these discourses challenged the prevailing order, Hajer's account indicates that neither became hegemonic. Instead, strands of these environmental discourses converged with the institutional practices of international policy-forming organisations, such as the OECD and UNEP in the formation of the ecological modernization discourse. According to Hajer, the historical roots of this discourse lie in this convergence between discursive and institutional forces, which rendered continued development compatible with environmental care (Hajer 1995: 101–2).

The discourse of ecological modernization acknowledges structural design faults within modern society that produce environmental problems but assumes that through modern political, economic and social institutional reform, care for the environment can be internalised (Hajer 1995: 25).[1] Thus storylines around ecological modernisation present environmental protection as a 'positive sum game', environmental degradation becomes a management issue and, in principle economic growth and ecological problems can be reconciled (Hajer 1995: 26).

[1] Ecological modernization was a concept first introduced by two German political scientists, Joseph Huber and Martin Janicke in the 1980s. Although originally an interpretation of how environmental policy had developed in Germany and the Netherlands (Langhelle 2000: 305), it is now deployed as both a social theory and a new policy-oriented discourse in environmental politics. For an introduction to ecological modernization and historical accounts of its development, see Buttel 2000; Fisher and Freudenburg 2001; Mol and Spaargaren 2000. Hajer's contribution is in tracing the emergence of the ecological modernization discourse during the 1980s and demonstrating its impact on the framing of the acid rain problem in the UK and the Netherlands.

Through his research of the acid rain problem, Hajer demonstrates that while different storylines competed to define the acid rain issue, dominant discourse coalitions formed around the ideas, concepts and categories of ecological modernization. Despite the success of the eco-modernist framing of acid rain, however, the institutional response was more consistent with prevailing end-of-pipe abatement strategies than the new policy discourse (Hajer 1995: 268).

The work of Karen Litfin and Maarten Hajer is important for disrupting dominant conceptions of science and politics, highlighting that the conceptualisation of international environmental problems is a contested arena. Litfin's empirical work identifies that scientists and scientific knowledge must compete with other actors in the construction of an issue for political response. Thus, far from remaining a separate and privileged domain, scientific knowledge becomes a force within on-going social and political struggles. Hajer's work illuminates how knowledge and social values are already configured in prevailing and emergent environmental discourses, which, along with existing national and international institutional structures, constrain the construction and treatment of a new environmental problem. Importantly, both discursive approaches make apparent that power is not concentrated in either scientific or political centres but rather is diffused across and between a range of actors in the problematisation process.

Although both discursive frameworks acknowledge social struggle, offer more diffused configurations of power and direct a critical gaze on a wider range of actors and institutional processes involved in the construction of environmental problems, neither Hajer nor Litfin systematically theorise the constitution and distribution of power between and across these actors and institutions. Questions remain over the distinct resources and forms of authority that each group of actor has in the struggle to define the problem, how authority is coupled to the distribution of material resources in broader social and political space and how this shapes how an issue is known addressed? The chapter turns to the theoretical approach and empirical insights of Steven Bernstein, whose work on the *Compromise of Liberal Environmentalism* sheds light on the power of ideas in global environmental governance and brings the importance of economic and political order into focus.

2.3 The Power of Ideas

Steven Bernstein's book on the *Compromise of Liberal Environmentalism* sets out to explain how 1970's environmentalism, premised on the incompatibility of environmental protection with socio-economic and political practices, evolved into liberal environmental governance, which predicates environmental protection on the promotion and maintenance of a liberal economic order. The norm-focused theoretical framework that Bernstein developed to explain this shift, illuminates some of the processes that constrain the definition of climate change as a social and political problem, which according to Bernstein's conclusions, must fit with the current economic order to be widely accepted. Importantly, this brings the relationship between environmental problem definition and patterns of broader political and economic order into focus.

Bernstein's constructivist explanation for the compromise of liberal environmentalism centres on his understanding of norms, norm-complexes and social structure (Bernstein 2000, 2001). Bernstein uses these to build a multi-layered normative framework to explain why some norms get selected over others and to explore the implications of this on governing global environmental problems. When new problems like climate change emerge, a space opens up for rethinking present governing structures and for criticising the social and political values that underpin these, as is characterised by the problematisation of environmental degradation in the 1970s. However, when these ideas attempt to become more than criticism and to initiate social and political change they are confronted with extant norm-complexes governing social and political relations. Bernstein suggests that these new ideas must 'compete against existing social purposes' and are only like to become institutionalised as governing norms by 'finding a fitness with those structures' (Bernstein 2001: 216). This 'fitness' is determined by: (1) the perceived legitimacy of the new ideas (who they came from and with what claim to authority); (2) the extent these new ideas fit with prevailing governing norms; and (3) the degree of fit with key actors' identities (Bernstein 2001: 184).

In the case of environmental protection, Bernstein finds that new ideas governing international environmental protection only became widely acceptable once they had found some fitness with norms of liberal economic growth and development. The notion of sustainable development then *is* the compromise of liberal environmentalism, legitimising a form of international environmental governance that 'predicates environmental protection on the promotion and maintenance of a liberal economic order' (Bernstein 2001: 213). As a result, the privatisation of the global commons and market mechanisms is not only perceived 'as compatible' with environmental protection but also necessary for successfully capturing and incorporating environmental concern in the practices of state and non-state actors. These conclusions have implications for understanding the processes by which environmental issues are conceptualised and defined as social and political problems. In contrast to Haas, who saw the direction of discovering, defining and understanding environmental problems proceeding from science to politics, Bernstein's approach suggests that the conceptualisation of environmental issues is not a linear process. It is not necessarily the case that problems like climate change are discovered, understood and defined by scientific communities before they can be governed, because current governing norms are likely to constrain the definition and framing of the issue before the physical extent of the problem is realised. Thus, although scientists may play an important role in raising the profile of the problem, once on the agenda, international political and economic structures shape policy-relevant research as much as vice-versa.

Evidence for this is in the extent that liberal economic theory and practice have impacted the institutional arrangements of the IPCC. Bernstein highlights how the content of Working Group III's (WGIII) report on climate change mitigation has increasingly focused on questions that fit with a liberal environmental research agenda. This is reflected in the increase in economists in the authorship and focus on cost-effective policy response options in the content since the second assessment report (SAR) (Bernstein 2001: 224–25; Corbera et al. 2016). This reorganisation of WGIII was driven by an attempt to enhance the political relevance of the IPCC's assessment for member governments (Bolin 2007: 80–81), rather than the result of scientific advances (Bernstein 2001: 171). It is the IPCC's assessment reports that have moved in line with a liberal economic order, and international political solutions are more likely to be accepted if they fit with norms of liberal environmental governance, as in the Kyoto Protocol, where reduction targets were linked to market mechanisms (Bernstein et al. 2010).[2]

Bernstein's work further complicates the relationship between scientific knowledge and political response. While from the discursive approaches we gained insight into a wider cast of actors, Bernstein's approach embeds the policy response in prevailing ideas of social, political and economic order. The effects of this can be traced through the IPCC's organisational development and observed in the research reviewed and promoted through the organisation's assessment activities. This mutual or co-construction between scientific knowledge and the political response to environmental degradation is further elaborated and tested by scholars of STS through the idiom of co-production

2.4 From Power to Social Order

Scholarship within STS has examined the intertwined relationship between natural and social orders in scientific knowledge of physical phenomena like climate change. As a discipline, STS brings to the fore – and essentially

[2] The three main market mechanisms within the Kyoto Protocol are: (1) Emissions trading; (2) Joint implementation (JI) among developed countries; and (3) Clean Development Mechanism (CDM). All three of the Kyoto mechanisms work on the same basic principle: 'that assigning property rights to emissions and creating a market that allows them to be transferred will enable emission reductions to be achieved where it is most cost efficient, or cheapest, to do so' (Bernstein 2001: 118).

accepts as its starting point – that scientific knowledge is deeply entangled with social norms and hierarchies (Jasanoff 2004a). Several STS concepts designed to unpack and characterise this entanglement have informed how the IPCC is understood and studied today, including boundary organisation and boundary work, as reviewed in Chapter 4. However, in terms of looking directly at the multiple ways that knowledge and order become imprinted on and internal to the construction of new environmental objects, it is the idiom of *co-production* that is most relevant (Jasanoff 2004a, 2004b; Miller 2004).

One of the central concerns of STS scholarship, as studied through the idiom of co-production, is the relationship between scientific and cultural processes through which new phenomena like climate change emerge and are stabilised as social and political objects. In this approach, as well as STS more broadly, science is not assumed to be an autonomous sphere 'whose norms are constituted independently from other forms of social activity' (Jasanoff 2004b: 30), but an activity whose connections with other social realms, such as the political, are to be studied. As Jasanoff elaborates in her discussion of co-production, the material and cultural resources through which actors bring new natural phenomena into view often exist 'before the "discovery" of the objects themselves' (Jasanoff 2004b: 16).

Jasanoff identifies institutions, and in particular the making of new institutions, as a central site of co-production (Jasanoff 2004b: 39–40). It is through institutions that societies have 'tried-and-true repertoires of problem-solving, including preferred forms of expertise, processes of enquiry, methods of securing credibility, and mechanisms for airing and managing dissent' (Jasanoff 2004b: 40). This approach again identifies the emergence of environmental problems as a source of new institutions, which 'emerge to provide the web of social and normative understandings within which new characterizations of nature … can be recognized and given political effect' (Jasanoff 2004b). The central addition to the previous approaches reviewed, however, is that the idiom of co-production brings the physical phenomena back into view. As Jasanoff states, 'it's not about ideas alone; it is equally about concrete, physical things' (Jasanoff 2004a: 6). It is the omission of the physical order in the social sciences that the idiom of co-production helps to guard against, reminding the researcher that explanatory power is gained when natural and social orders are thought about 'as being produced together' (Jasanoff 2004a: 2).

Putting the idiom of co-production into practice, Clark Miller (2004) develops an alternative account of the politicisation of climate change, highlighting the power of ideas and suggesting that the climate had to be re-constructed as a global phenomenon before its significance as an international political issue could be recognised. Miller describes how the scientific framing of climate change was dramatically recast between the early 1960s and the late 1980s, transforming from a concept that was used to describe local and regional long-term weather patterns to one that depicted a global-scale phenomena. He concludes that the re-imaging of the climate as a global system was central to the production of climate change as an object of international political action, as it 'brought views of the atmosphere in line with assumptions about the jurisdictions of international organizations' (Miller 2004: 51).

Miller's and other STS informed accounts identify the ascendency of Global Climate Modelling (GCM) as central to the transformation of climate change into a political object (see also Shackley and Wynne 1995; Demeritt 2001; Edwards 2010; Allan 2017). The computer modelling of the general circulation of the atmosphere meant that scientists increasingly studied and represented climate change as an integrated global system – a cognitive framing that was clarified and extended through the establishment of the IPCC and its first assessment report (FAR). Miller suggests that by bringing 'concepts of natural order and political order into line with one another, the IPCC served to co-produce new arrangements of global nature and civil society' (Miller 2004: 55–56). For this reason, Miller identifies the IPCC as both a product and an agent of co-production and stresses that the utility of the co-productionist idiom is that it attunes the scholar to 'the multiple ways that knowledge and order become coupled in the emergence of new objects like climate change' (Miller 2004: 61).

2.5 Summing Up

This chapter has sought to explore the different models, frameworks and approaches that scholars have developed to study the emergence of an environmental problem and the scientific and political processes through which these issues are known and addressed. Conceptions of scientific knowledge evolved through the review of the chapter, from the view of science and politics as separate spheres to the notion of knowledge as a resource for political actors to mobilise and acknowledgment of the intertwined nature of scientific knowledge production with social and political orders. These evolving accounts of knowledge identify the importance of convergence between the issue and the existing order in how an environmental problem like climate change is known and addressed.

The discursive and ideational-based accounts demonstrate that knowledge of issues like climate change is structured by existing institutionalised political and economic norms and practices. Miller's account, for example, highlights that cognitive frames of climate change had to resonate with and connect to the institutional remit of international organisations to become a recognisable object of international political action. Bernstein reaches a similar conclusion when he finds that new ideas and new issues, like climate change, had to find some fitness with the prevailing liberal economic order. This economic order of relations not only pre-existed the discovery of climate change but is ultimately productive of the problem and its continuation is ensured through the (neo)liberal economic response measures that are prescribed as the solution. Nearly all the accounts reviewed stressed that the emergence of an environmental issue is frequently the source of new institutions. However, here too Hajer's conclusions are cautionary, highlighting the continuation of existing order through the practical response. As a result, despite discursive challenges to present ways of thinking and doing, institutional practices can remain unchanged in the practical implementation of a response.

These accounts of the necessity of convergence between the problem and present ways of knowing and doing change the narrative of the IPCC's emergence. The chapter began recounting the story of the IPCC's emergence through epistemic community accounts. These identified the scientific workshops and events that served to constitute an epistemic community on climate change and create the avenues through which growing scientific understanding of climate change and its consequences was communicated to policymakers. However, Miller's account suggests that it was not just a case of increasing scientific knowledge and its communication, but a shift in how climate change was known that is also key to understanding the establishment of the IPCC. It was through the emergence of climate change as a global phenomenon that required a collective response, which aligned the issue with the purview of international organisations. Bernstein's account also highlights the work that was necessary to create this alignment and to maintain political interest in the IPCC, which increasingly turned to economists and economic methods for evaluating policies to mitigate climate change.

In terms of addressing the study of order and power asymmetries in the construction of environmental problems, however, there remain conceptual and methodological gaps. Although Jasanoff indicates the potential of a co-productionist approach to account for social hierarchy, the forms of order that Miller's narrative highlight are globalising social and institutional arrangements. This leaves social order as a distribution of distinct forms of power within and across actors unaddressed. As I explore who has the power to construct climate change and what constitutes this power, there are two analytical capabilities that the book needs. The first is a way to locate the IPCC in broader political space, to ensure that the social order studied within the IPCC is situated within global distributions of economic, social and political resources. The second is to identify and measure the distribution of resources that is constitutive of the social order within the IPCC. It is this situating of the IPCC in global climate politics that I turn to next.

3

Situating the IPCC as a Practice of Writing

It was after my seventeenth interview that analytically something shifted. It was an uncomfortable interview, where the participant felt defensive, and it reflected the atmosphere for the Intergovernmental Panel on Climate Change (IPCC) at the time. When I asked about the target audience and the direction of the next assessment, the respondent impatiently reminded me that 'What you should do is really read the rules and procedures we have to follow' (interview 5.10.2002). I had read those rules and procedures; I had been referred to them in another awkward interview with a government delegate. I just could not seem to grasp the content of this document the way my respondents did. The discomfort of this interview turned out to be a critical moment on my journey into the IPCC, and my understanding of what this organisation is and does in relation to climate change began to form.

A few days after this interview, I travelled to Busan in South Korea to observe the 32nd plenary of the IPCC, which enabled me to watch how the rules and procedures were written. The year 2010 was a difficult one for the IPCC. It was subject to fierce criticism when errors were found in WGII's assessment of the likely melting and disappearance of the Himalayan Glaciers, the so-called 'Himalayagate' affair (Carrington 2010). This came on the back of a wave of scepticism that followed the hacking and publication of email exchanges between prominent climate scientists and IPCC authors from the University of East Anglia in the run up to COP 15 in Copenhagen in 2009, coined 'Climategate' (Pearce 2010). To address these criticisms and re-establish the organisations symbolic power, the UN Secretary-General, Ban Ki-moon, and IPCC chair, Rajendra Pachauri, requested the IAC 'to conduct an independent review of IPCC processes and procedures used to produce assessments (IAC 2010a, 7). It was at the 32nd plenary that the IAC review and recommendations were discussed by the panel, and the processes and procedures for producing assessment reports were reformulated.

It was on observing the rewriting of the rules and procedures that I came to understand why interview participants kept referring me to this document.[1] The rules and procedures are a reference to practice – describing and prescribing how IPCC assessment reports are to be constructed, by whom and performing which tasks. Organisational documents such as these embody IPCC and plenary specific activities that put them together.[2] As scholars have indicated in relation to other organisations, to be understood as documents embodying practice, they need to be observed in the making, which makes apparent that the institutional processes through which they are constructed are as important as their contents.[3] Within the context of the plenary sessions, these documents become rich texts, packed full of signposts and references to the social forces and dynamics that govern the interactions between actors during the meetings.

It was through observation that IPCC documentation came to life. Suddenly, I felt that I could read the IPCC paper trail and 'get' these documents as my participants did. It was at this point that I saw the significance of the assessment report's construction pathway, and it became clear that how an IPCC assessment report is assembled – its journey through the IPCC, who this provides access to and authorises to perform set activities – is central to the meaning imbued in the document through the process. From this, I came to share scholar's perspective on the importance of the view from the inside for understanding the making of organisational documentation (Riles 2000; Neumann 2002; Hull 2003, 2012; Yamin and Depledge 2004).

I made another important observation at this meeting, something I knew on paper, but which I observed in practice. Not all actors are equal in the construction of IPCC documents; there is a pecking order (Pouliot 2016) in and to their writing. Put simply, there are those actors that speak and shape the construction of the document versus those that speak but the text remains unchanged. The politics and power on display at this meeting were palpable and entangled. There was the phenomenon of the disinterested country delegate, the effect of which was empty spaces behind country plaques and people wandering in and out of the hall during proceedings. In contrast, the deeply immersed and invested delegations dominated the proceedings, evidenced by the number and timing of interventions recorded in Chapter 4 (Table 4.2). On the surface these appear as two distinct

[1] In relation to the UNFCCC, Yamin and Depledge (2004: 2) state that 'documents alone give little insight into the functioning of the regime because it is difficult to glean the institutional practices, procedures and informal understandings that help define how the international climate process actually works.' Other scholars studying UN documentation have made a similar observation, see for example, Rise (2001).

[2] As noted by Yamin and Depledge (2004: 470), 'In some cases, the IPCC has developed informal customary practices that, while well established, are not recorded in the text.' This explains why, for many interviewees, the questions I asked seemed self-evident, but to someone not familiar with IPCC customs, the documents and procedures prescribed were often impenetrable.

[3] Hull observes the same in his study of document-making in bureaucratic institutions in Pakistan, 'Things look different from the inside' (Hull 2003: 289).

and disconnected phenomena. However, a connection becomes apparent through asking, what constitutes the power of some actors to speak, be heard and shape the text, while the words of others are lost in proceedings?

Critical to illuminating the social order within the IPCC is understanding where the organisation is located in political space – its place in climate politics. When analysed, the IPCC is often situated to the side of climate politics, as a provider of knowledge for action rather than as a site of political action in and of itself. In this chapter and inspired by the sociology of Pierre Bourdieu, I redescribe climate politics as a struggle to fix the meaning of climate change and thereby how it is acted on and the order that those actions make or retain (Bourdieu 1991). This situates the IPCC centrally in climate politics as a powerful site of order-making in the struggle to name the problem. Symbolic power has emerged as an important dimension of Bourdieu-inspired study of power and authority in IR (Epstein 2008; Stuvøy 2010; Abrahamsen and Williams 2011; Adler-Nissen 2013; Eagleton-Pierce 2013; Adler-Nissen 2014; Adler-Nissen and Pouliot 2014; Hughes 2015, 2023; Sending 2015; Hughes and Vadrot 2019). However, its imprint on organisational products is less frequently subject to systematic analysis and the aim of the next two chapters is to develop an analytical framework that can enable this. Using Bourdieu's notion of field to situate the IPCC in political space is essential for identifying the external pressures and forces this social location generates and examining how these dynamics have historically shaped, and continue to shape, the IPCC and its assessment practice. From this location in climate politics, I move inward to the IPCC's organisational form, and by the end of the chapter, I begin to put together the analytical framework that will enable me to identify the actors, activities and forms of authority that constitute the IPCC's practice of writing climate change.

3.1 The Politics of Naming Climate Change

Thinking of climate politics as a struggle to name climate change helps to make sense of this complex realm and sensitises study to the forces it generates. Climate change is recognised as a super wicked problem that confounds conventional ways of knowing and responding to collective issues (Levin et al. 2009, 2012). At a societal level, the battle to contain and control this problem has played out through contestation over the human role in and physical extent of climate change. Within the climate negotiations, from the outset the struggle has been to determine who acts to reduce greenhouse gases (GHGs) (Paterson and Grubb 1992), when and by how much alongside how and by whom this mitigation, adaptation and loss and damage are to be funded (Rajamani 2015a, 2015b, 2016). The notion of naming, however, points to the forces and dynamics that lie beneath and run through these visible struggles; it points to the implications that the meaning of climate change has for international order and ultimately for life itself. Naming climate change offers the potential for world-making on an unprecedented scale. The way this issue is constructed shapes the societal response at all levels, which has the potential to reconfigure all practices of life as these are impacted by and adapted to the shifts and extremes of a changing climate and aligned with carbon-neutral social organisation.

While the daily lives of all are likely to be impacted by physical changes and the political response – regardless of responsibility for the problem – not all have the power to name the problem. The stakes in this struggle are indeterminately high. For those privileged within the current fossil-fuel dependent global order, there is social dominance to preserve and maintain. For others, there is the necessity to attain some level of this 'development' and the opportunity to advance a global political order based upon and organised around value for human and environmental relations. These stakes have engendered a 30-year fight over climate change that continues to intensify, and the IPCC is a central battle ground. To explore the effects of this struggle on the organisation, its assessment practice and the knowledge produced, the first step is to situate the IPCC within this global struggle to fix the meaning of climate change and the field of activity this has generated.

The condition of a field is interest, as interest and investment is what produces struggle and generates the forces that structure a field (Bourdieu 1986a; Wacquant 1989; Bourdieu and Wacquant 1992).[4] The emergence of interest in climate change has a long history, although until the 1980s, this was largely an object of scientific enquiry, to be discovered and represented through scientific practice. As described in Chapter 2, during the 1980s a group of actors, workshop and assessment activities brought climate change to the attention of a wider audience. The political interest these actors and activities generated were critical to the formation of the IPCC and for transforming climate change into an object of wider social concern (Hughes 2015).

When the IPCC was established in 1988, its mandate was to assess the most up-to-date knowledge of the science, impacts and response strategies to climate change (UNGA RES/45/53, 1988). This established the IPCC as a central site for naming climate change and, as such, placed the organisation and its assessment practice in the middle of emerging political interest in the issue and the forces and struggles this generated. However, once political interest was mobilised, and

[4] For other studies that have adapted Bourdieu's concept of field for the study of organisations and 'international' objects, see Bigo 2006, 2011; Fligsten and McAdam 2011, 2014; Pouliot and Mérand 2012; Sending 2015. The focus here is on exploring the emergence of interest in climate change and to identify some of the key forces and dynamics that come to structure relations and the newly forming institutions established to address climate change (Hughes 2015), in other words, the genesis of the field (see Sending 2015).

as the stakes in the issue became ever more apparent, the global response to climate change took on a momentum that was not within the organisation's design or power to control. While actors within the parent bodies of the United Nations Environment Programme (UNEP) and World Meteorological Society (WMO) envisioned that they would carry forward the process from the IPCC and initiate negotiations for a framework convention on climate change – as UNEP had for ozone depletion – political contestation made this impossible.

A UN General Assembly resolution supported a UNEP decision to begin preparations for climate negotiations in December 1989. However, when an open-ended ad-hoc working group of government representatives was convened by UNEP and WMO in September 1990, disagreement quickly arose. Participants at the meeting could not agree on who should organise and conduct negotiations: 'a negotiating committee under the auspices of WMO and UNEP, in essence carrying forward the IPCC process, or a special conference under the authority of the UN General Assembly' (Bodansky 1993: 473–74). While countries in the global north generally supported the former option, many countries in the global south, 'who felt excluded from the IPCC, preferred the second option' (Bodansky 1993: 474).[5] As a result, an International Negotiating Committee (INC) under the auspices of the General Assembly was established for negotiating the framework for collective action on climate change (UNGA RES 45/212 1990). Although this newly formed body was to 'take into account' the work of the IPCC, and UNEP and WMO were invited to make 'appropriate contributions' to the process, a separate ad hoc secretariat was established (UNGA RES 45/212 1990).

The formation of the INC transferred the IPCC's mandate for formulating policy response options to this newly formed body (UNGA RES 45/212 1990). The 1992 adoption of the UN Framework Convention on Climate Change (UNFCCC) established this treaty-making process as the principal site for negotiating global interest in climate change. However, like the IPCC, the UNFCCC does not determine the limits of the climate field nor is it able to retain control over the outcomes of its negotiating processes. The social, scientific, political and economic ways of practicing climate change generated through the UNFCCC – temperature targets, methodologies for measuring and reporting, financial transactions through mitigation, adaptation and loss and damage funds and the new institutional arrangements that oversee and implement these – do not stay in the hands of those that authorise their creation. In fact, as methods for embedding climate change in the everyday organisation and conduct of life, these objects come to exert their own force both on the UNFCCC process and on the wider climate field.[6]

[5] See also Miller 2001a: 255, 2004:59–61.

[6] See for instance, Paterson (2009) on the 'quasi-autonomous dynamic' of the 'carbon market' and Bernstein et al. (2010) on how these markets have taken on a life of their own beyond the reach of states to control them.

With the creation of the negotiation site under the auspices of the United Nations General Assembly (UNGA), the organisation of climate change knowledge for these negotiations became more deeply separated than it might have been had both organisations been under UNEP.[7] This meant that the managers of the IPCC had to undertake work to define and maintain the organisation's role as the central knowledge provider (Hughes 2015: 95). During the negotiation of the Framework Convention, the IPCC chair and other actors within the organisation sought to establish the IPCC's institutional significance to this new body. However, this objective was hampered by a number of developing countries, most notably Brazil and India, which had limited authorship in the first assessment report (FAR) (IPCC 1990a: xxviii), did not accept the assessment's construction of climate change (IPCC 1990b: 151), and did not want the IPCC formally recognised within the Convention text (Miller 2001b: 255; Biermann 2002: 205–6; Yamin and Depledge 2004: 465).

Consequently, the IPCC's attempt to be signified as the official provider of scientific and technical advice was unsuccessful and a provision was made for the establishment of a Subsidiary Body for Scientific and Technical Advice (SBSTA). This new body was to manage the Convention's knowledge requirements with only oblique reference to relations with 'existing competent international bodies' made in the text (UNFCCC 1992, Article 9).[8] The SBSTA was not formally constituted, however, until the first meeting of the Conference of the Parties (COP 1) in 1995. In the meantime, the IPCC was requested to respond to the Conventions needs for 'objective scientific and technical advice' (UNFCCC 1992, Article 21). This gave the IPCC leadership time to strengthen the organisation's position as leading knowledge provider in the climate field and to establish its relationship to the UNFCCC.[9]

The IPCC's success in establishing its relationship to the UNFCCC is apparent. The SBSTA and COP regularly request the IPCC to provide assessments and input specific information into the negotiating process (IPCC 2007a), and both the Kyoto Protocol and the Paris Agreement refer to the IPCC and its work (UNFCCC 1997, 2015). The IPCC's relationship to the UNFCCC has been further consolidated by the Paris Agreement. In the Paris Agreement, the IPCC continues to provide the methodology for Parties to provide regular reporting on anthropogenic emissions of GHG

[7] This was the case, for example, with negotiations on ozone, the Basel Convention and Convention for Biological Diversity.

[8] For a detailed analysis of SBSTA and its relation vis-à-vis the UNFCCC and the IPCC, see Miller 2001a, 2001b.

[9] The decisions taken at the first Conference of the Parties in 1995 strengthened the IPCCs formal tie to the UNFCCC, 'citing it as a source of "the latest international scientific, technical, socio-economic and other information", as well as input on methodological issues' (Yamin and Depledge 2004: 465). The relationship between the IPCC and the SBSTA was also formalised, with the SBSTA charged with summarising and converting the IPCC's assessments into a format 'appropriate to the needs of the Conference of the Parties' and seeking input and advice from the IPCC on methodologies, for example, for compiling national GHG inventories etc. (IPCC 2007b: 3).

by sources and removals by sinks (UNFCCC 2015, Article 13), as it did in the Kyoto Protocol (UNFCCC 1997, Article 5). A role that is extended to providing the methodologies and common metrics for reporting in Nationally Determined Contributions (UNFCCC 2015, decision 1/CP.21). Furthermore, the IPCC assessments are identified as a source of input to the global stocktake, and the organisation was invited to provide a special report in 2018 on the impacts of global warming of 1.5°C above pre-industrial levels (UNFCCC 2015, decision 1/CP.21). As explored in Chapter 7, this privileged relationship means that many of the key objects negotiated within the UNFCCC, including the 1.5°C temperature target, originate from or are legitimated through the IPCC's assessment practice (Fogel 2005; Lahn and Sundqvist 2017; Livingston and Rummukainen 2020; Beek et al. 2022; Cointe and Guillemot 2023). This relationship to and role in global climate agreement-making generates forces that require constant attention and management by lead IPCC actors.

The IPCC's position in the climate field rests upon its symbolic power to arbitrate over the legitimate means for knowing and acting upon climate change, a source of power that is coupled to its relationship with the UNFCCC. The IPCC competes for this position with other climate change knowledge and assessment products compiled by national governments, NGOs and other international organisations, and maintaining its relationship to the negotiating process is crucial for preserving the relevance and pre-eminence of its assessment products. This pre-eminence means that the IPCC and its assessment practice have the capacity to authorise climate knowledges and expertise and thereby legitimate actors' stake in international climate politics. Governments are well aware of the force that the IPCC's knowledge production has in and over the negotiating process and for this reason take a keen interest in the IPCC's procedures for producing assessments and approving the report outline (Chapter 5) and key messages in the SPM (Chapter 7). This political interest exerts a force over the IPCC's practice of writing and the content of the final messages, which is documented over the coming chapters. However, it is not only governments that seek to gain access and input into the IPCC's climate change assessment.

Global interest in climate change diverts attention and resources from other international political concerns, including global health issues (Fidler 2010), migration (Hall 2015), biodiversity (Jinnah 2011a), desertification (Conliffe 2011) and other environmental problems (Axelrod 2011). If these issue areas are to retain the interest of the international community, they must either recapture attention from climate change or align themselves with and to the interests of the climate field.[10] In the case of global health, the IPCC assessment process

[10] The strategic alignment of issue areas with the climate regime has been conceptualised as climate bandwagoning; see the special issue of Global Environmental Politics (Jinnah 2011b).

has provided an important means for identifying the impact of climate change to human health (Hashimoto et al. 1990; McMichael et al. 1996, 2001; Confalonieri et al. 2007) and for highlighting the synergies between tackling climate change alongside other development issues (Field et al. 2012). The primary concern and objective of actors within the global health field may be to recapture political commitment and resources to long-standing global health issues. However, by becoming interested and invested in climate change and supporting research on the health impacts of climate change, the field of health can align their interests with the interests of the climate field and promote their expertise and the utility of their work through the position they take (Bowen et al. 2012). This also brings important benefits to the IPCC and supporters of the UNFCCC process, as health issues mobilise public and government support for addressing climate change, which in turn deepens support for further research and assessment activities (Haines et al. 2007, 2009; Haines 2008). This demonstrates how interest and investment in climate change shapes both the climate field and other fields in diverse ways: diverting energy and attention from other global concerns. At the same time, the IPCC's assessment practice provides an opportunity for actors to develop new forms of capital and revenue in their own fields, as well as empowering their form of expertise within the climate field.

The IPCC's capacity to legitimate knowledge makes it the prime target for those contesting the science of climate change and the political significance of the issue. The notion of naming identifies this as a struggle over social order, with actors contesting the reality of climate change and the knowledge that underpins it to preserve their own social position (Lahsen 2008). In an attempt to undermine the authority of the IPCC's scientific findings and contest the reality and urgency of climate change, actors have sought to undermine the credentials of IPCC authors and the scientific conventions of its assessment practice (Edwards and Schneider 2001; Dunlap and McCright 2015; BBC Four 2021). By publicly exposing inconsistencies in the process and content of a new report, these attacks have successfully challenged key findings that underpin the collective response, forcing the IPCC to defend its work and moderate its operating procedures to regain authority for the next assessment. The measures the organisation takes to re-establish its authority are critical. If governments and other users of the reports come to question the IPCC as a legitimate source of knowledge, or if they suspect that the IPCC is no longer widely recognised as a legitimate source of knowledge, they may want to reduce their proximity to the organisation to avoid becoming embroiled in the controversy and preserve their own symbolic power.

The IPCC came under sustained attack prior to and after COP 15 in Copenhagen in 2009 as a result of the 'climategate' affair and mistakes discovered in the fourth assessment report (AR4).[11] The IPCC chair was targeted in these attacks and strongly criticised for his response to errors over the melting of the Himalayan glaciers.[12] As the face of the organisation, calling into question the ability of the IPCC chair poses a threat to the organisation's symbolic power. Traditionally, the IPCC chair addresses the UNFCCC COP during its high-level segment. However, the IPCC was not invited to present to the main plenary at COP 17 in Durban in 2011 (Gutierrez, van Alstine and Yamineva 2011: 8). This provoked consternation amongst IPCC panel members during the 34th plenary session, and after informal communication between the IPCC Chair and the UNFCCC Executive Secretary, the agenda was amended (Gutierrez, van Alstine and Yamineva 2011). Although these events may be unrelated, the criticism directed at the IPCC chair and its reports impacts the organisation's scientific authority and its position in the climate field, which has the potential to impact upon those that are closely coupled to the organisation and its assessments. It is therefore unsurprising that, at a time when the UNFCCC was dealing with questions around its own centrality to climate action (Keohane and Victor 2011), the secretariat sought to insulate itself by distancing itself from the IPCC and its chair (Hughes 2015: 97).

This section has set out to describe the IPCC's role in establishing global interest in climate change, to locate the IPCC within the climate field and to sketch the dynamics that this social location engenders. Recounting this history reveals that neither the IPCC nor the UNFCCC have been able to contain interest or contestation over climate change, which is revealing of the physical phenomenon and collective effort to reach agreement on it. By implicating the current economic order, climate change threatens current ways of organising life and those privileged within and by this order. At the same time, by bringing modes of existence into focus, climate change offers an unprecedented opportunity for re-evaluating the basis and values of collective organisation. The improvement this could bring to the material conditions of some at social and economic cost to others makes the stakes in this struggle indeterminately high. The physical reality of climate change is increasingly encroaching on the organisation and conduct of everyday life. As the consequences of a changing climate become everyday experiences, the limits of the climate field will continue to expand with the potential for every being to become interested and invested in its name. The IPCC is centrally situated in this expanding social field, and this growing interest exerts pressure on what the IPCC is, the activities that identify it as such and the meaning these activities produce. These external forces, however, are not the

[11] On 'Climategate' see Pearce 2010. On mistakes surrounding the melting of Himalayan glaciers, see Carrington 2010.

[12] For information on this attack on the IPCC chair, see Section 5.3.

sole pressure on the IPCC and its assessment practice; the IPCC's mandated purpose and the necessity of realising global reports generate their own pressures. To identify these internal forces, I turn from the IPCC's location in the climate field to the IPCC as its own field of practice.

3.2 The IPCC as a Field of Practice

The climate field is a field in emergence. As a field orientated around activity in the name of climate change, it is shaped by actors from well-established fields of professional activity. Originally limited to scientific practice, today global interest in climate change is generated by and constitutes political, scientific, economic, legal and bureaucratic activities. Actors draw their authority to know and respond to climate change, and symbolic power to influence and be influential in the climate field, from these diverse disciplinary and professional realms. Each has distinct interests in the problem, ways of practicing the profession, recognising authority and undertaking activity in and for the name of climate change. Collectively, over time, however, these distinct ways of doing and knowing professional existence are developing internationally recognised ways of practicing the climate change problematic, which in turn shapes the fields contributing to this collective attempt at agreement-making.

Like the climate field in which it is situated, the IPCC is made up of actors from distinct fields of practice, from the scientists that produce knowledge of climate change to the member governments that accept and approve IPCC products. Each of these fields is driven by its own interests and has its own conventions and methods for producing and recognising legitimate ways of knowing, as well as means for assigning and acknowledging authority. Over the past 30 years, actors' shared investment in realising an intergovernmental assessment of climate change has produced a unique organisation and practice for producing assessment reports. Participants' investment in the organisation and its assessment activities has exerted a force that shapes actor interests over time and through practice. It is this reshaping of participants through their involvement in the IPCC that enables the organisation itself to be identified as a field of practice.

Mapping the IPCC as a practice of writing enables a detailed study of by whom (the actors), how (through what activities) and on the basis of what authority the reality of climate change is written, and the order of relations this practice is built upon and has the potential to remake. As I aim to establish over the coming chapters by borrowing Bourdieu's notion of capital (Bourdieu 1986b), the IPCC embodies the international order of which it is the product, an order that has the potential to be reproduced in and through the IPCC's practice for producing knowledge of climate change. That is, if the IPCC's practice of writing remains static and uncontested. It is the IPCC's symbolic power to write reality, however, which makes its practice of writing climate change – by whom, on the basis of what authority and according to what scholarly, administrative and diplomatic conventions – a central object of struggle within (and outside) of the organisation.

Producing internationally recognised assessments of climate change that can be accepted and approved by member governments is a delicate balance and the product of struggle between scientific conventions, political interests, administrative realities and the social and economic structures these embody. Objectively, in and through the structures of the world, the struggle for symbolic power in and over the IPCC's practice of writing is fought over recognised pillars of the organisation: rules and procedures for producing IPCC assessment reports; geographical and disciplinary representation in the authorship and leadership of the assessments; access to and distribution of labour in writing the reports. Subjectively, these battles are governed by the perceptions, opinions and attitudes of IPCC participants, as internalised in habitus. Less often cited, these cultural dispositions are a conservative force, as they have a capacity – despite organisational attempts to ensure balance – to recognise, acknowledge and reproduce the order of which they are the product (Hughes and Paterson 2017).

The production of IPCC assessment reports began as a relatively informal process, one that relied on the conventions and experiences of the actors leading the process. However, as the political stakes in the climate change problematic increased and as the IPCC leadership sought to be at the centre of the international political response, pressure on the rules and procedures for constructing IPCC assessment reports also increased. Political involvement combined with relentless criticism about the inclusiveness and scientific authority of IPCC assessment products persistently brings the IPCC's rules and procedures into focus, which has resulted in the codification of the IPCC's practice of writing. This practice is the product of the interplay between external pressures and the internal struggle between scientific, political and administrative authority in and over how an IPCC assessment is to be written, by whom, on the basis of what qualifications and according to what geographical balance.

Geographical representation of the IPCC's panel, bureau and authorship has been a force within the IPCC's practice of writing from the outset (Agrawala 1998a, 1998b), the significance of which is demonstrated by the contestation of developing countries of the IPCC's position and symbolic power in the climate field (Hughes 2015). Objectively, the IPCC's struggle to legitimate and universalise its assessment products has played out through the creation of funds for developing country participation (Chapter 6), bureau expansion and author numbers (Chapter 6). Subjectively, this is a harder battle, as in practice scientific and other cultural criteria identify and distinguish leadership through the assessment process and these forms of authority are tied to the material necessities required for building a distinguished scientific career (Chapter 6).

Access to the IPCC's practice of writing is power. This is as true for those within the IPCC as it is for those attempting to gain access to the climate field and legitimate their stake in the struggle over climate change. The objective of the IPCC's practice of writing is an intergovernmental assessment of climate change – knowledge that ultimately has the symbolic power to challenge and legitimise particular ways of perceiving and acting on the world in the name of climate change. Access to the assessment's assembly pathway and the activities through which the reports are compiled – outlining, commenting, assessing, reviewing, editing, selecting the report's core messages and accepting and approving the final product – offers the opportunity to shape how climate change is known. Those invested in the IPCC's practice of writing struggle for and over the forms of capital that enable increased access to and influence over the writing of climate change. Interests in and objectives for access depend on actors' role and position within the organisation and are largely the product of the scientific, governmental and bureaucratic fields that qualify actors to participate in the assessment process. This individual investment in climate change and the IPCC cannot be reduced solely to a search for recognition (Sending 2015), as this sits alongside a deep-seated care for human and planetary relations that the IPCC transforms into meaningful and purposeful action.

The organisational form and assessment practice of the IPCC are products of these struggles. Today the IPCC can be identified as five distinct units according to actors' role, interests and authority in and over the assessment practice. These five distinct units have emerged over the course of the IPCC's lifetime and in response to the forces exerted on the organisation and its practice of producing assessments of climate change. The following chapter identifies these units as the panel, the bureau, the WG, the technical support units (TSUs), the secretariat and the authors. Chapter 4 sketches the relations between these units, the activities they perform and the access and forms of capital they have in and over the IPCC's practice of writing. Once the organisation of the IPCC has been mapped, it is possible to follow the production of an IPCC assessment report along its assembly pathway, which is divided into the decision to repeat the assessment (Chapter 5), the scientific assessment (Chapter 6) and the acceptance and approval of the final product (Chapter 7). This is a living organisation and a dynamic process that is constantly updating and adjusting to the demands and forces of its situation, not least the force that a changing climate exerts, and the practice of writing aims to capture the nature of this process rather than produce a fixed representation of its outcomes.

3.3 The Method for Unravelling a Practice

The IPCC's assessment practice is made up of a multitude of tasks – scheduling, arranging, meeting, reading, reviewing, writing, compiling, editing, commenting, submitting, intervening, proposing – activities that are divided between set actors that have responsibility for and the recognised authority to conduct these tasks and differentiated capacity to influence through the process. To systematise the study of how these activities enable access to and authority over the IPCC's assessment practice and the distribution of symbolic power to write climate change through the process, I disaggregate the organisation according to actors, activities and forms of authority. Adopting actors, activities and forms of authority as an analytical framework makes it possible to unravel the IPCC's practice of writing into its constituent tasks and to describe the emergence of this practice over time and six rounds of assessments. This brings to light the properties or forms of capital that have been valued in the organisation and the social order of relations this has institutionalised within the IPCC.

This approach offers new insight to studies of the IPCC and has the potential to open up other international organisations to systematic study because no particular group of actors or forms of authority are privileged prior to study. Thus, while existing studies of the IPCC informed by the concepts of epistemic community, notions of co-production or boundary organisation (as reviewed in Chapter 5) focus analysis on actors and relations designated as scientific or political, this categorisation is not reified from the outset because all actors that are part of the organisation are included in the analysis. This brings in to view the secretariat and technical support units, which have previously been overlooked and under analysed. In Chapter 5, I describe the particular forms of authority housed within these units and identify who within the organisation has privileged access to the capital these actors hold. In this section, I explore how this analytical approach emerged through the fieldwork and identify the forms of data collected and used to build an account of the IPCC as a practice of writing climate change.

Looking back, there were three distinct stages to my research that contributed to reconstructing the IPCC as a practice of writing (see Table 3.1). In reality, this has been a process of back and forth, for example, returning to IPCC documentation after an interview or revising interview questions based on observation. First, I focused on the assessment reports, recording the nationality, expertise and career trajectory of the authors, and how this changed over subsequent assessments. I read the executive summaries of the chapters, the technical summaries and SPMs of each round of assessment reports to try and get a sense of how climate change was being constructed, through what forms of knowledge, and how this shifted over

Table 3.1 *Methodology for studying the IPCC as a practice of writing*

Analytical framework	The practice of writing 1. Actors: the actors that make up the IPCC – grouped into units according to the activities they undertake and the forms of authority this gives in/over the assessment 2. Activities: the set tasks that a unit has/is responsible for in the production of the report 3. Authority: the valued properties and their distribution within and across units of the organisation that enables actors to shape the conduct and content of the report
Method	1. Documentary analysis: Identification and location of actors; description of IPCC assessment activities; organisational concerns; change over time 2. Interviews: bibliographical/historical account of interest in climate change; roles in the IPCC; description of activities; perceptions, attitudes, dispositions and values (who or what is valued) 3. Observation: place and type of work; role in the organisation; social order – who has an effect; struggles and contestation
Aim	1. Reconstruct the IPCC's assessment practice 2. Identify the actors with symbolic power, the properties and attributes of this power to imprint and its distribution at different stages of the assessment formation 3. Identify the attitudes and dispositions shaping the social order within the IPCC and its potential imprint on IPCC products 4. Explore the relationship between valued social properties within IPCC (scientific authority) and global political order (economic resources)

time through each round of assessment. This initial research was important for identifying key actors within the organisation to interview.

I was interested in speaking to long-standing IPCC participants and those that had contributed to multiple assessments in order to document how the assessment process had developed over time. By the time of writing, I had conducted over forty interviews with IPCC authors, bureau members, government delegates and TSU staff (see appendix 1), as well as informal conversations and email exchanges over a ten-year period. Roughly half of these were face-to-face interviews that took place in the UK and North America and averaged just over an hour in length. One quarter of respondents were interviewed in their place of work, which included a visit to the WGII TSU at the Carnegie Institute at Stanford University in 2010 and a visit to the WGIII TSU at Imperial College London in 2019. I transcribed all interviews in full and used this process to reflect on my interview style and the material I was gathering. After the first two or three interviews, I began to recognise respondents accounts from the published literature, and I became aware that respondents were contributing to and influenced by IPCC scholarship. To get beyond shared opinions on the IPCC, I changed my interview questions, asking interview respondents to recount the origins of their professional interest in climate change, their initial encounters with the IPCC and then to detail the tasks they undertook as a participant.

The change in interview questions made it possible to build a picture of the daily working environment of IPCC participants, which was critical for identifying and locating actors within a field of expertise and comparing and contrasting this to their activities in the assessment process. During the interview stage, the opportunity arose to attend and observe the proceedings of the 32nd IPCC plenary session, hosted in Busan, South Korea, in October 2010, which became the third stage of data collection.[13] IPCC plenary meetings are the annual meetings of the panel that are organised by the secretariat and attended by IPCC member governments, the bureau, TSU staff and organisations with observer status. This four-day long intergovernmental meeting was the most significant stage of the research process for two reasons. First, I observed the distinct way each unit has for conducting its business and how these are adjusted when the organisation works together in performing a joint task, as it does during plenary proceedings. Second, it gave me an insight into the construction of IPCC documentation and the practical purpose this construction process and its end products serve.

The final account of the IPCC as an organisation and an assessment practice is provided by reference to IPCC documentation, the InterAcademy Council investigation into the organisation (IAC 2010a, 2010b), Earth Negotiation Bulletins of IPCC plenary sessions, secondary literature and where gaps remained, further correspondence and follow-up interviews. In most instances, I do not directly quote interview respondents in the text. I used interviews to understand how the IPCC worked and to gain detailed accounts of assessment activities, rather than as a means to gather information on any particular person or event. The aim is to reconstruct a collective process – a shared social practice – which could not be built from the perceptions or point of view of a single actor. For this reason, all interview data was cross-referenced with other sources, including IPCC documentation, previous historical accounts of the IPCC and the IAC (2010b)

[13] I obtained observer status through the Tyndall Centre, and I am indebted to Mike Hulme and Asher Minns for enabling this.

questionnaire. Table 3.1 indicates how each method of data collection focused on particular knowledges and understandings of the IPCC that enable it to be constructed as a practice of writing.

3.4 Summing Up

In this chapter, I have described climate politics as a struggle to name the problem driven by attempts to determine the basis by which climate change will be named and the global order of relations this will preserve or has the potential to remake. The IPCC, as the recognised knowledge provider for collective action, is situated centrally within this struggle – authorised to write and set the rules and procedures by which the global meaning of climate change is to be written in and through its assessment practice. The IPCC's practice for writing climate change is shaped by the political forces and struggles that the emergence of political interest in climate change engendered and the organisational attempts to contain and respond to these over thirty years and through six rounds of assessments. The pathway for producing assessments that has emerged – the IPCC's practice of and for writing climate change – can be broken down and studied into its constituent parts through the actors, activities and forms of authority that constitute it. In Chapter 4 I use actors, activities and forms of authority as the analytical framework to disaggregate and to describe the social order that structures the organisation, its conduct and its products.

Examining the IPCC through this analytical framework identifies the IPCC as five distinct organisational units: the secretariat, the panel, the bureau, TSUs and authors. The units are distinguishable by the actors and by the distinct sets of tasks that actors within each unit are authorised to undertake in the writing of climate change. If you remember, in concluding Chapter 2, I suggested that there were two things that I needed from the analytical approach of the book. The first was to situate the IPCC within climate politics so that it was possible to discern the impact of this position on the organisation and its assessment practice, which the chapter did through re-describing climate politics as an act of naming. The second, was to enable an exploration of the authority to name – the properties and distribution of this symbolic power to designate the reality of the problem, and its relationship to broader patterns of global economic and political order. In the coming chapters, I describe the three key stages in the production of an IPCC report: the outline (Chapter 5), the assessment (Chapter 6) and the approval of its key messages (Chapter 7). The analytical framework will aid in providing an account of the order of relations that govern the IPCC's practice of writing through each of these stages and enable exploration of whether this order is written into or challenged in the naming of climate change through the process.

4

Analysing the IPCC as Actors, Activities and Forms of Authority

It was after a day of interviewing in July 2010. I was waiting for my dinner, and it suddenly dawned on me; I don't know what I am studying. What is the IPCC? I had just interviewed a member of Working Group I's (WGIs) technical support unit (TSU) for the third assessment report (TAR). I struggled to keep up. She was both a scientist describing her research on climate change and a member of the TSU. In the TSU, she was scheduling the timeline of the assessment, facilitating meetings, supporting the chair and, at the same time, an author in the technical summary and 'making sure the science is ok' (interview 07.07.2010a). She was performing multiple roles in the assessment and described an organisation that I had not grasped from the literature. I decided then and there, as I scribbled on the menu, that one thing I've got to do is accurately describe this organisation. This is the aim of the chapter. I reconstruct the IPCC by exploring its historical emergence through the actors, activities and forms of authority that constitute its present form, as I learned it through descriptive interviews and later, personal observation.

The IPCC has been of particular interest to scholars within IR, STS and human geography. Despite differences in disciplinary approaches, two core concerns are shared within the literature. The first is the relationship between science and politics: how this shapes the organisation and its knowledge products and informs the collective response. The second is participation and particularly its diversity in terms of gender, academic disciplines, forms of knowledge and critically to this study, asymmetries between developed and developing country participation. Many of the studies reviewed in this chapter provide historical accounts of the organisation's establishment and identify the forces that have shaped its development. However, in focusing on scientific and political relations and dynamics within the IPCC and its work, they overlook aspects of the organisation that could not necessarily be identified as either, such as the TSU. Applying the framework of the book makes it possible to take the IPCC apart as an organisation and to identify and describe all actors, activities and forms of authority regardless of whether they could be labelled as scientific or political. Doing this enables the chapter to build on the study of asymmetries in knowledge production and its effects, not just within the authorship of reports as previous studies provide, but in the everyday conduct of the organisation and its practice of writing climate change.

4.1 Constructions of Science and Politics in the IPCC

Established in 1988 under the co-sponsorship of WMO and UNEP, the IPCC was mandated to assess the science of climate change, its social and economic impacts and policy response strategies (UNGA res 43/53). This task was divided between three Working Groups (WG), each charged with assessing a dimension of the climate change problematic: the physical scientific basis of climate change (WGI); the impacts of climate change (WGII); and response options (WGIII). This remit was adjusted after the first assessment report (FAR), when the responsibility for formulating policy response options was transferred to the newly established International Negotiating Committee (INC) and WGIII was re-focused to assess 'cross-cutting economic issues' (IPCC 1992a, 14). While the remit of the WGs has adjusted with advances in understanding and the needs of its users, the focus on producing comprehensive assessments has remained constant over 30 years and 6 rounds of assessment reports. These assessment reports, along with special and methodological reports and the accompanying technical summary and summary for policymakers (SPM), provide governments with the accepted knowledge base for negotiating climate action within the UNFCCC. It is the IPCC's role in informing the global community's response to climate change that has made the organisation a key site for studying science in politics.

As reviewed in Chapter 2, the early origins and historical development of the IPCC are conceived as the successful outcome of an epistemic community of climate scientists, performing their function by bringing the causes and consequences of climate change to the attention of policymakers and framing the issue for collective debate (Lunde 1991; Boehmer-Christiansen 1994a, 1994b; Paterson 1996; Haas 2000; Newell 2000). The establishment of the IPCC in 1988 is viewed as the product of this successful politicisation process (Paterson 1996). Despite the success of the IPCC's intergovernmental process in raising the political profile of climate change and initiating an international negotiating process, scholars from within IR in particular have been critical of the political involvement of member governments in the IPCC, questioning the organisation's capacity to function as an information or scientific advisory institution to the climate change regime (Biermann 1999, 2002; Haas 2000, 2004; Haas and McCabe 2001; Haas and Stevens 2011; Stavins 2014). It is the intergovernmental nature of the IPCC and the proximity between science and politics this creates, which makes the IPCC and its assessment products an ideal site for unpacking how science informs politics and how both are shaped through these interactions.

Early contributions to this body of research originate from the Global Environmental Assessment (GEA) project (Parson et al. 1997; Mitchell et al. 2006).[1] This multidisciplinary group of scholars expanded knowledge of assessment activities, conceived as 'the entire social process' through which expert knowledge of a particular issue 'is organised, evaluated, integrated and presented in documents to inform policy or decision-making' (Parson et al. 1997: 53). The GEA's conception of assessments incorporates both the 'products and reports, and the process that generates them' (Parson et al. 1997), with the aim of understanding how organised scientific efforts shape societal response to global environmental change. The project's workshops and fellowships generated some of the most detailed historical accounts of the IPCC's establishment (Agrawala 1998a, 1998b; Biermann 2000, 2002; Miller 2004). These empirical studies illuminate the role that organisations like the IPCC perform in meditating between science and politics in producing credible knowledge, suggesting that success is reliant upon participant's ability to maintain the distinction or a 'boundary' between the worlds of science and policy in and through the assessment process (Guston 1999, 2001).

As study of the IPCC has matured, 'boundary organisation' and 'boundary work' have emerged as central concepts for characterising the IPCC and its assessment activities, with the IPCC identified as 'the most significant' or 'preeminent' boundary organisation on climate change (Adler and Hirsch Hadon 2014: 663; O'Neill et al. 2015: 380; Beck et al. 2016).[2] The metaphor of boundary has brought insights on how science and politics are intertwined in the organisation and how this intertwinement shapes final knowledge products (Shackley and Wynne 1996; Shaw 2000, 2005; Skodvin 2000a; Miller 2001b, 2004; Siebenhüner 2003; Fogel 2005; Petersen 2006; Lövbrand 2007; Hoppe, Wesselink, and Cairns 2013; Beck et al. 2016; Lahn and Sundqvist 2017; Beck and Mahony 2018; Gustafsson 2019; Livingston and Rummukainen 2020). One of the most revealing studies in this regard is Clark Miller's analysis of the Subsidiary Body for Scientific and Technological Advice (SBSTA).[3] Miller identifies the institutional arrangements

[1] The GEA project was a multidisciplinary project launched in 1995 to address questions concerning the ways in which organised efforts in scientific information provision shape social responses to large-scale global environmental change. Many of the scholars referenced in the following section participated.

[2] The term 'boundary organisation' and the notion of 'boundary work' were developed in the GEA project by the scholarship of David Guston (1999, 2001) influenced by Thomas Gieryn 1983.

[3] The SBSTA is a permanent subsidiary body to the UNFCCC, providing scientific and technological advice on matters related to the convention to the Conference of the Parties. For more information, see: https://unfccc.int/process/bodies/subsidiary-bodies/sbsta (last accessed 29 March 2023).

within the SBSTA and the IPCC as 'amalgamations of social practice drawn from the worlds of both science and politics', rather than two distinct domains as they may appear on the surface or as claimed by those within them (Miller 2001b, 483). Miller uses *hybrid* to refer to institutions that house these amalgamations of practice, suggesting that to maintain a productive relationship, boundary organisations like the IPCC need to be able to manage hybrids ' – that is, to put scientific and political elements together, take them apart, establish and maintain boundaries between different forms of life, and coordinate activities taking place in multiple domains' (Miller 2001b, 487).

One of the ways this management of hybrids has been studied is through the concept of boundary work, which enables exploration of the intertwinement of science and politics in practice. Fogel, for example, describes how both authors and member governments undertake *boundary work* during the approval of IPCC reports when they assert claims to science through notions of objectivity, bias and policy prescriptiveness in their attempts to maintain or revise the content (Fogel 2005). This has led scholars to conceptualise IPCC products, particularly elements that travel to or emerge from the negotiating process as 'boundary objects' (Lahn and Sundqvist 2017; Livingston and Rummukainen 2020; Beek et al. 2022; Lahn 2022). These studies reveal that while the organisation's authority may in part rest on maintaining and promoting the demarcation between science and politics, the relevance of its products requires drawing these worlds together and making concrete connections across the two domains (Lahn 2022).

This scholarship provides rich empirical accounts of the ongoing and negotiated relationship between science and politics in the IPCC and its products. However, scientific and political actors, activities and forms of authority are not the only social dynamics and structuring forces that shape the IPCC or its practice of writing. The TSUs have greater day-to-day contact with the assessment than any other actor, and yet the activities they undertake and the forms of authority this gives them over the assessment are not accounted for in this scholarship. To identify the valued properties – or forms of authority – that shape the conduct and content of the assessment and their distribution, in Section 4.2, I return to the historical emergence of the organisation when its cultural foundations were laid. This makes it possible to explore a second core concern in IPCC scholarship, namely the asymmetries in developed and developing country participation.

Developing country participation has been an issue for the IPCC since its establishment (IPCC 1988, 1989; Schneider 1991; Sagar and Kandlikar 1997; Biermann 1999, 2000, 2002; Kandlikar and Sagar 1999; Siebenhuner 2002; Gupta 2013; Hughes 2015, 2023; Okereke 2017; Yamineva 2017). Although the IPCC has supported some developing country participation at IPCC meetings since 1989 (IPCC 1990a) and made mandatory support for at least one developing country author on every chapter since the second assessment report (SAR) (Agrawala 1998b), research continues to identify the extent that economic and cultural barriers impede research and authorship from the global south. Studies on research expenditure and its link to output reveal how economic resources, as measured by Gross Domestic Product (GDP), impede developing countries from generating their own climate knowledge and related expertise (Ho-Lem et al. 2011; Pasgaard and Strange 2013; Livingston et al. 2016). The dominance of the English language is a compounding factor, impacting the peer-reviewed literature assessed as well as non-native speaker communication within chapter teams (Ho-Lem et al. 2011).

The literature has highlighted how domestic forces also shape developing country contribution to the IPCC's assessment practice. Historically emerging out of scientific interest in Europe, the international climate research agenda has been led by the investment and research interests of the US and the UK (Kellogg 1987; Hart and Victor 1993; Hecht and Tirpak 1995; Edwards and Lahsen 1999; Shackley 1999). Scientific interest in climate change was not matched in developing countries, where other pressing social and environmental issues commanded the attention of researchers and the limited resources of governments. For instance, when the IPCC was established in 1988, scientific efforts in India were concentrated on local pollution issues that were considered of greater social and political relevance (Kandlikar and Sagar 1999), with a similar situation in Brazil (Lahsen 2004: 167). Even a decade after the IPCC's establishment, Indian government funding agencies did not give as much value to lead authorship in the IPCC reports compared to North America and Europe, which limits the career benefits for authors contributing to this time-consuming process (Biermann 1999; Mahony 2014:113–14). As Borland, Morrell and Watson's (2018) study of one climate research centre in South Africa highlights, the constraints of limited resources have to be seen in combination with conscious decision-making to invest in local development priorities, industry partnerships and policymaking. As a result, the national context may not place as much value on international journal publications as a measure of contribution and scientific authority as the social order within IPCC author teams (Corbera et al. 2016; Hughes and Paterson 2017).

Sociological study into the global economy of knowledge reveals that southern countries that have emerged as important climate knowledge producers, such as Brazil, continue to remain dependent on theories, techniques and models developed in the global north (Connell et al. 2018a, 2018b). The result is 'asymmetrical partnerships' in North-South research collaboration, with Southern partners valued as local experts rather than as co-producers in theoretical and methodological problem construction (Connell et al. 2018b: 5–8). Taken together, the impact of these disparities on IPCC assessments is threefold: first there remains a lack of data for some of the most climate vulnerable regions in the world (IPCC 2007b; Pasgaard and Strange 2013; Pasgaard et al. 2015; Livingston, Lövbrand and Olsson 2018). Second, developing country authors are perceived as less credible scientific contributors to the assessments and have less influence over the content (Corbera et al. 2016 Hughes and Paterson 2017). Third, developing county knowledge, perspectives and concerns on issues that have profound effects on domestic and global climate decision-making are not adequately represented in scientific literature and IPCC assessments of this (Kouw and Petersen 2018; Livingston, Lövbrand and Olsson 2018; Biermann and Möller 2019).

IPCC scholarship discerns how disparities between research expenditure, interest and investment impact on developing country authorship and the content of the final assessment. However, work remains to be done to unpack how and to what extent economic and cultural properties also structure the broader social order of the IPCC and the everyday conduct of producing a report. To what extent do measures of and asymmetries in knowledge production and scientific authority impact relations in the panel, in the bureau and in the administrative and technical support of the assessment and with what effect for developing country participation? To explore this systematically, there is a need to disaggregate the different sets of actors that make up the IPCC, to describe the activities they undertake and to identify the distinct forms of authority this gives them in and over the assessment.

4.2 The Units of the IPCC

The remainder of this chapter aims to describe the IPCC as it has historically emerged as an organisation in its current form, which can be identified as five distinct units: the panel, the bureau, the TSUs, the secretariat and the authors (see Figure 4.1). For the majority of the actors that make up these units, IPCC activities are not a full-time occupation: it is an author's contribution to climate change knowledge from within the field of science and a delegate's position within the meteorological office or ministry of the environment that authorises actors to participate in the IPCC's assessment practice. However, over time and through participation, distinct scientific, diplomatic and administrative ways of practicing knowledge production have developed a shared way of realising an intergovernmental assessment of climate change. In achieving this, activities and authorities have been divided, distributed and struggled over within and between these units as each attempts to access, gain authority and increase their symbolic power in and over the IPCC's practice of writing. It is the combination between the imperative of realising the mandate, internal social dynamics and the external forces generated from its central location in climate politics that structure the organisation and its assessment practice today as described in remainder of the book.

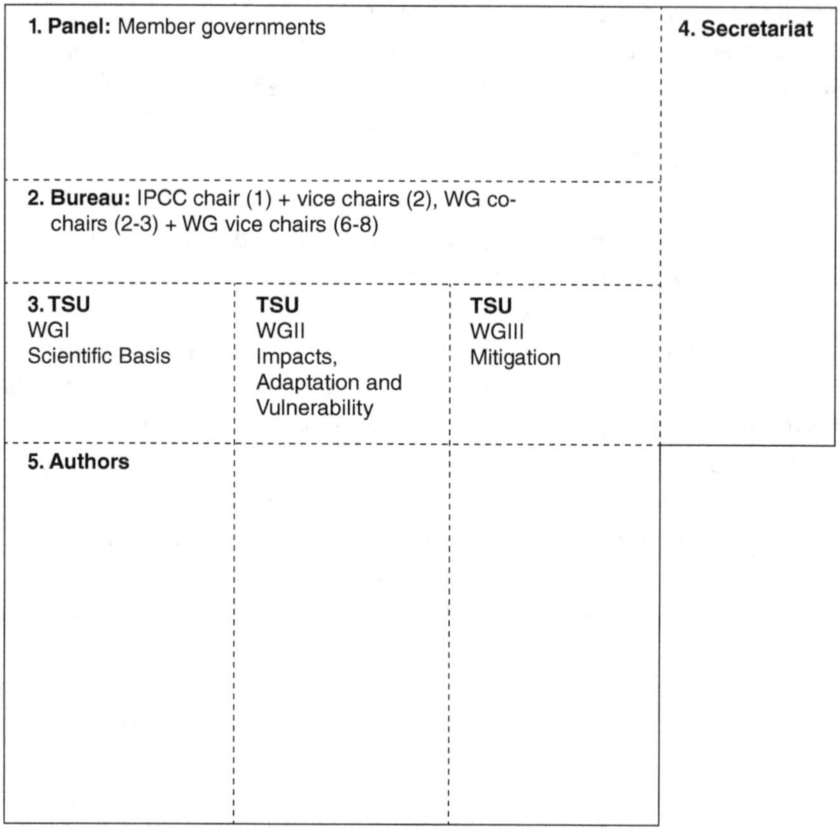

Figure 4.1 The IPCC represented as five distinct units: (1) panel; (2) bureau; (3) secretariat; (4) TSUs; and (5) authors. Units 1–4 come together for the IPCC plenary and have access and share information with each other. WG bureau and TSU have direct contact with the authors. First published in Hughes 2023. This schematic does not include the Task Force on National Greenhouse Gas Inventories (TFI).

4.2.1 The Panel

The Panel is the IPCC's member governments that meet once or twice a year in plenary session (see Figure 4.2). Membership to the panel is open to all member countries of WMO and UNEP and there are currently 195 members (IPCC n.d.). However, only half regularly send representatives to plenary, and for reasons unpacked later, about one quarter could be described as engaged in panel activities (IPCC 2009a). The panel is involved at every stage of the IPCC's assessment practice, apart from the authorship of the report, and governments

Figure 4.2 The panel: Government delegates seated in alphabetical order at the 40th plenary of the IPCC (Copenhagen, Denmark, 24–31 October 2014). Photo by IISD/ENB: https://enb.iisd.org/climate/ipcc40/.

have considerable influence over the organisation and its work (Hughes 2022). Although member governments are not directly involved in the authorship of assessments, governments approve the report outline, nominate authors, elect the bureau, review draft reports and accept and approve the final products (see Table 4.3). Financially, the IPCC is dependent on donations, and all IPCC expenditure is agreed by the panel, which gives governments the final decision over the organisation's continuation, its assessment activities and the expert meetings and workshops supporting these.

The nationally designated focal point is usually based within the meteorological office or environment and related departments and acts as an intermediary or conduit between the IPCC, the government and the national scientific community. Between plenary meetings, this actor is engaged intermittently in IPCC work, overseeing the national process for identifying and nominating authors, managing the government review of draft reports and preparing for plenary and approval meetings (IPCC 2010d). To become a meaningful member of the panel, governments must invest the economic and human resources necessary to fulfil this broad range of activities, and through the conduct of these, governments can gain authority in the organisation and influence over the direction and content of the report. This includes, through the bureau election and approval of the next report outline (Chapter 5), nominating authors and submitting review comments (Chapter 6), and interventions, requests and red lines during the approval of the key findings of the reports (Chapter 7). This symbolic power in and over the IPCC's practice of writing is not equally distributed between member governments, with historical involvement and, relatedly, knowledge of the process being the most important forms of cultural capital structuring relations within the panel.

4.2.1.1 Historical Involvement

Those countries that played a leading role in the founding of the IPCC and in the production of its FAR lay the cultural foundations of the IPCC's assessment practice, which distinguished the valued properties within the panel. The IPCC's FAR was originally envisioned as an exercise for a small group of core members and although all WMO and UNEP members were invited to the IPCC's first plenary session, only 30 countries sent delegates (IPCC 1988; interview with BM 1.07.2010). Of the 90 government representatives at this meeting, 61% came from 8 countries: US (15); Japan (11); Canada (8); USSR (5); Australia (4); France (4); Netherlands (4) and; UK (4). The first IPCC chair, the Swedish scientist Bert Bolin, recognized the importance of broadening geographical participation and in 1989 the panel set up a special committee to assess and increase developing country representation (IPCC 1989; Bolin 2007: 55). By this time, however, the mandates and work plan of the three working groups (WG) had been established and the principles for their operation formulated, principles that would later be codified and adopted by the panel as the rules and procedures governing IPCC assessment activities.

The current modus operandi of the IPCC is underpinned by the valued properties and principles of its founding members. As documented by the Australian delegate to the first IPCC session and later a long-standing bureau member, Australia's 'emphasis on the importance of objectivity, the involvement of subject matter experts and the use of peer-review procedures during its interventions at the first session, significantly shaped the character of the IPCC in its early years' (Zillman 2007: 873), and it was on US insistence that peer-review was incorporated into the assessment practice (Zillman 2007). The acceptance of these scientific principles without debate (Bolin 2007: 49–52) indicates the shared nature of scientific practice and corresponding cultural values within and between US, Canada, Northern Europe and Australia who were leading the process. This embodied working style, or habitus, was infused into IPCC proceedings, as summarised in Bolin's address to the panel, in which he urged members' decision-making to be grounded in scientific and technical arguments:

> He reminded the panel that the IPCC is not a negotiating body.... He hoped there would not be much need for decision-making by voting in the IPCC.... In this process, it was most important that the developing countries were given adequate opportunity to take part because the process then led to mutual learning, benefitting not only the developing countries but also the developed countries.... So orderly conduct of business in a free and scientific manner with participation by all or as many as possible should be the IPCC working mode. (Paraphrased in report of the session, IPCC 1991: 6–7)

This statement designates an appropriate style of conduct, privileging scientific and technical forms of argumentation in panel discussions and automatically empowering those embodying this way of operating. This contrasts with the view from developing country delegates, many of whom identified gaps in national data and scientific capacity (IPCC 1991) and felt a 'sense of frustration' in the process because of the human resources required (in speech by Mostafa Tolba, IPCC 1991: 5). Many developing countries contested the scientific and technical framing of climate change and called for the issue to be recognized as a developmental issue and be assessed in the context of sustainable development (see Zimbabwe speech to first session in IPCC 1998 annex 3, 11; Borione and Ripert 1994: 81). Thus, while membership to the panel rapidly expanded from 1988 onwards (Agrawala 1998b), and in theory it is possible for all member governments to gain or increase influence in and over assessment activities, countries must possess the economic capital to invest in participation and act according to the cultural mode embodied in the IPCC's practice of writing.

4.2.1.2 Knowledge of the Process

Knowledge of the process is a valuable source of cultural capital within the IPCC (see Table 4.1). There are several avenues through which to accrue knowledge of the process, including length of participation, involvement in panel activities and having a national expert elected as a bureau member. Governments with an elected bureau member can attend bureau meetings and draw on this contact within and between sessions, which provides these governments with greater insight into the assessment process. This knowledge of the process is a valuable source of cultural capital in the IPCC – translating into symbolic power during plenaries and report approval sessions when delegates can draw on their insider perspectives to make informed interventions and authoritative reasons for altering proposed text.[4] As Table 4.2 indicates, those countries intervening most in plenary proceedings all have bureau members. This relationship is strongest

[4] Joanna Depledge (2007) uses the term intellectual capital to identify the experience and knowledge that UNFCCC Secretariat and chairpersons have and its value to other actors (also Bauer 2006, Jinnah 2010). Intellectual capital can be identified with both Bourdieu's notions of cultural capital (knowledge, skills, technical qualifications and titles) and social capital, in the sense that it is only a source of capital to those that have a connection/relationship to those in the know and thus a pathway to access it.

Table 4.1 *The activities and forms of authority of the panel*

Actor	Activities	Forms of authority
Panel Member governments represented by delegates Currently 195 members (IPCC n.d.) Usually reside within department for environment/ climate change or national meteorological organisation	• Decision to produce report • Financial expenditure • Approve the report outline • Nominate authors • Elect the bureau • Review and submit comments on draft reports • Accept final report • Line-by-line approval of report outline and assessment's key findings in SPM • Focal point: intermediary between national expert community, national government and the IPCC	*Economic capital* • Government investment *Cultural capital* • **Political authority** • Historical involvement/ length of service • Knowledge of the assessment process • Knowledge of IPCC rules and procedures • Knowledge of assessment content • Hosting TSU • Bureau membership • Authors in the assessment *Social capital* • Bureau member • Hosting TSU • Relationship with secretariat • Relationship with/ to other member governments

Table 4.2 *Top ten countries by frequency and total time of interventions at the 32nd Plenary Session of the IPCC, hosted in South Korea, October 2010 (data collected by author)*[1]

Top country by interventions (Department listed for focal point in 2023, IPCC n.d.)	Number of interventions	Top country by total time of interventions	Total time (seconds)
1. US* (WGII) Department of State	50	1. Switzerland* (WGI) Federal Office for Environment	4,849
2. Switzerland* (WGI) Federal Office for Environment	43	2. US* (WGII) Department of State	4,240
3. Saudi Arabia* Ministry of Petroleum and Mineral Resources	33	3. Saudi Arabia* Ministry of Petroleum and Mineral Resources	3,218

4. Australia* Department of Climate Change, Energy, Environment and Water	28	4. Australia* Department of Climate Change, Energy, Environment and Water	2,854
5. UK* Department of Business, Energy and Industrial Strategy	25	5. UK* Department of Business, Energy and Industrial Strategy	1,960
6. Belgium* Science Policy Office	24	6. Russia* Institute of Global Climate and Ecology	1,532
7. Germany* (WGIII) Federal Foreign Office	24	7. Netherlands Infrastructure and the Environment	1,288
8. Netherlands Ministry of Economic Affairs and Climate Policy	23	8. Germany* (WGIII) Federal Foreign Office	1,222
9. Austria Federal Ministry Agriculture, Forestry	14	9. Austria Federal Ministry Agriculture, Forestry	1,062
10. Sweden Meteorological and Hydrological Institute	12	10. Brazil* Ministry of Foreign Affairs	942
Totals	**276/433**		**23,167/33,431**

* Member countries with a bureau member.
[1] Only interventions from the floor, and not presentations by delegates or bureau members chairing contact groups, were included.

where a developed country co-chairs the WG, and hosts the TSU, with these governments intervening most frequently during plenary sessions. The US delegation is symbolically powerful in this regard. Actors from within the United States played a critical role in establishing the IPCC and laying its cultural foundations (Hecht and Tirpak 1995; Agrawala 1998a, 1998b), and the United States hosted a WG chair and TSU for five consecutive assessments (see table 4.4). This role in the establishment of the organisation, chairing of WG assessments, and hosting of TSUs has ensured that US cultural ideals and principles of scientific legitimacy, expertise and authority have underpinned the development of the IPCC's practice of writing and the representations it generates.

In contrast, the majority of developing country members have struggled to acquire the necessary capital to act as authoritative participants in the panel. An important factor in understanding this is the organisation of participation domestically. The national focal point acts as an intermediary between the government, national scientific community and the IPCC and the location of this actor within the government bureaucracy matters. While 32% (59/185) of national focal points to the IPCC are located in the meteorological department (IPCC n.d., IPCC focal points), out of the 10-member governments that took up 69% of the airtime at the 32nd plenary (Table 4.3), today only 1 of these continues to have a focal point located in the meteorological department, Sweden.

In part, the location of the focal point is a marker of the recognition given to the IPCC and the climate issue domestically as well as how international relations are organised in the country. When the IPCC was established, most governments coordinated their participation through the meteorological or related department. However, as the salience of climate change has increased, coordination has tended to move to environment-focused departments, with some country particularities, such as the US, where national participation to international organisations is coordinated by the State Department. The location of the focal point impacts on the coordination and degree of investment in IPCC activities (interviews 17.09.2010; 5.10.2010), such as the identification of experts for author nomination, who within the government (if anyone) participates in the government review of assessment reports and the relaying of information between the IPCC and UNFCCC processes. Interviews with past and present focal points reveal the level of investment that countries, including Australia, United Kingdom and the United States, put into coordinating IPCC-related activities (interviews 26.07.2010; 13.12.2010; 13.12.2010), which translates into symbolic power during the approval of IPCC products.

4.3 The Bureau

The IPCC bureau oversees and manages the production of IPCC assessment reports and in this function is an intermediary between the member governments of the panel that authorise the assessment and the expert authors that produce the report. In relation to the panel, the bureau's main purpose is to provide scientific and technical advice to guide member governments (IPCC 2011, see Table 4.3), and the opinion of the bureau has a significant influence on panel decision-making. Today, the bureau and the panel are distinct units within the IPCC, each with specific functions and forms of authority over the assessment process. This distinction between the bureau and panel has developed over time and in response to pressures from within and outside of the organisation.

As indicated earlier, the IPCC's establishment was led by a relatively small group of individuals identified as representatives of government, the parent

Table 4.3 *The activities and forms of authority of the bureau*

Actor	Activities	Forms of authority
Bureau 34 members for the AR6: IPCC chair, three vice chairs, two co-chairs and seven or eight vice-chairs for each WG (IPCC n.d.) Scientific experts nominated and elected by member governments of the panel Usually reside in a university, research institute or relevant government department	• Provide scientific and technical advice to the panel to support decision-making WG Co-chairs • Oversee and manage the assessment • Select authors • Chair approval of outline and final SPM WG Vice-chairs • Support WG co-chairs in above roles • Identify and mobilise regional expertise • May act as review editors or on cross-cutting issues across chapters and WGs	*Economic capital* • Government or institutional support • Trust fund for developing country travel *Cultural capital* • **Scientific authority** • Scientific reputation (contribution to science/ publications + institutional affiliation) • Historical involvement • Experience of international scientific proceses and assessment exercises • Knowledge of the assessment process • Necessity/centrality to completing assessment, for example, CLA *Social capital* • TSU • National focal point • National delegation • Scientific/professional networks

organisations (WMO and UNEP) and/or prominent members of the international climate science community. The original structure and work programme of the IPCC was largely agreed between these actors before the first IPCC session and accepted during the plenary without much debate (IPCC 1988; Bolin 2007: 49–50; Zillman 2007: 872). A bureau of fifteen was appointed to oversee the work of the FAR at the first session, which included the IPCC chair, a vice-chair, a rapporteur and a chair and vice-chair for each of the three WGs (IPCC 1988).[5] It was decided that to 'provide for the best possible co-ordination' that appointed bureau members should be, 'where possible, Principle Delegates of their respective countries in IPCC' (IPCC 1988: 6). This indicates the fine line between the bureau and the panel at the time of the IPCC's establishment. However,

[5] Except for WG III, which because of governments' interest in the assessment of policy response options had five vice-chairs. For more details, see Bolin 2007 49–52; IPCC 1988. For an account of how US government departments vied for chairmanship of the three working groups see Hecht and Tirpak 1995.

Table 4.4 *Countries that have hosted TSUs by WG and assessment round*

	FAR (1990)	SAR (1995)	TAR (2001)	AR4 (2007)	AR5 (2014/5)	AR6 (2022/3)
WGI Science	UK	UK	UK	US	Switzerland	France
WGII Impacts	USSR/ Australia	US	US	UK	US	Germany
WGIII Mitigation	US	Canada	Netherlands	Netherlands	Germany	UK

as political interest in climate change has increased, so has the government's interest in the work of the IPCC and the desire for greater autonomy over the organisation and its assessment activities.

To ensure regional representation, today's bureau has 34 members representing the 6 WMO regions, plus additional representatives from Africa and Asia.[6] The line between bureau and panel membership is today distinct, with few of today's national delegates serving on the bureau and vice versa. Some perceive that this has led to a loss of authority for bureau members in relation to the panel (interview 13.02.2010), while others consider it a necessary development (interview 9.11.2010). As described earlier, key bureau members, particularly the IPCC chair, played a central role in instilling the cultural values of the international scientific habitus in plenary proceedings, which distinguished the value of scientific and technical forms of authority in the order of relations.[7] This means that bureau members were able to deploy scientific conventions and measures of authority for the purpose of containing and channelling political forces during decision making and approval of text. However, over time, panel members have become resistant to these challenges, seeking to instil a more familiar negotiating style, which is oftentimes judged as confrontational and obstructionist by bureau members.

The development of cultural contestation between the bureau and the panel over the conduct of IPCC proceedings is epitomised in the controversy surrounding the election of the IPCC chair, which also illustrates these competing cultural forces.[8] In 2002, the incumbent IPCC chair, Dr Robert Watson, was not re-elected for a second term in an election process that divided opinion within

[6] It also includes the two co-chairs of the Task Force on National Greenhouse Gas Inventories.

[7] As John Zillman (2007: 873) indicates: 'In the initial stages of its work, the Panel operated essentially according to the General Regulations and meeting procedures of the WMO with its Vice-Chair, two of its three Working Group Chairs and several other members of its Bureau all being experienced in the WMO and/ or UNEP systems.'

[8] For more on 'cultural contestation' and how it develops between different units of an organisation, see Barnett and Finnemore (1999: 724).

the panel (Lawler 2002; Zillman 2007: 875).[9] This was the first time in the IPCC's history that it was necessary for the panel to take a vote on the position of chair, a decision that until then was reached by acclamation between panel and bureau members. This precedence and the lack of codified procedure for its resolution further complicated the process. Although the IPCC chair is said to be stateless, the two most commonly cited explanations are US opposition to Watson's re-election and the necessity of the chair position to be held by a developing country member (interview 13.12.2010; Bolin 2007: 185–87). The United States supported the election of Dr Rajendra Pachauri from India, who unlike his predecessors was a WGIII expert with a PhD in industrial engineering and economics. There was a perception amongst some within and outside of the IPCC that WGIII did not do real science, which reveals the cultural hierarchisation of knowledges.[10] Pachauri was neither recognised for his contribution to climate science or for work on earlier international environmental processes, which automatically called his credibility into question (Lawler 2002).[11] US support for Pachauri's election was perceived by some as an attempt to undermine the scientific authority of the IPCC, or at the very least to contain its influence over the climate field at a time when the administration was hostile to the UNFCCC negotiating process and its Kyoto Protocol (Haas 2004).[12]

Despite the panel's attempt to demarcate itself from the bureau and the cultural contestation between the panel and the bureau to determine the order of proceedings, the panel remains reliant on the bureau to oversee the production of IPCC assessment reports. The developed country co-chairs are amongst the most powerful actors in this regard. Recognised for a combination of scientific contribution and experience of similar international environmental processes, the WG co-chairs are responsible for the management and production of the

[9] Watson was elected chair in 1996 by the US government, his re-election was opposed by the State department, which is said to have been under pressure from the Council on Environment Quality during the Bush administration (McRight and Dunlap 2010: 120).

[10] Evidence of this is recorded in Lahsen's (2008) sociological analysis of physicists' attitudes towards climate science and climate scientists and Shackley's quote of one of the climate sceptics that Lahsen's study analyses: '… why are the opinions of scientists sought regardless of their field of expertise? Biologists and physicians are rarely asked to endorse some theory in high energy physics. Apparently when one comes to "global warming" any scientist's agreement will do' (Richard Lindzen (1992) quoted in Shackley 1996: 204).

[11] In contrast, Bert Bolin the first chairman of the IPCC (1988–1997) had a PhD in meteorology, contributed to knowledge on carbon-cycle science and was a central actor in the establishment of a number of international research programs. He led a UNEP, WMO and SCOPE assessment of climate change (SCOPE 1986), and also acted as advisor on science policy to the prime minister of Sweden (Rodhe 1991). Sir Robert Watson (IPCC chairman 1997–2002) has a PhD in Chemistry. Prior to his chairmanship of the IPCC he chaired the Scientific and Technical Advisory Panel to the Global Environment Facility (1991–1994) and became Senior Scientific Advisor to the World Bank's Environment Department in 1996. He was also the associate director for Environment in the Office of Science and Technology Policy in the Executive Office of the President in the White House.

[12] Pachauri identified his election as 'a mandate for his plan to emphasize the socioeconomic effects of climate change on specific regions of the world' (in Lawler 2002).

WG assessment. The WG co-chairs are contracted by the national government for a percentage of their time to work on the IPCC process, alongside technical and administrative support in the form of a TSU, which is housed in or near the co-chair's institution, as described in Section 4.4. The developed country co-chair leads every stage of the report's compilation: drawing up the report outline (Chapter 5); selecting the authors (Chapter 6); overseeing the assessment (Chapter 6); preparing the summary for policymakers (Chapter 7); and chairing the approval of this document (Chapter 7) (see Table 4.4). The WG vice-chairs assist the co-chairs in this role, and the degree to which the vice-chairs and developing country co-chairs imprint their expertise on the process depends on their scientific credentials, experience of assessment processes and the extent to which they invest themselves in IPCC work, with considerable variation noted by bureau members during interviews.

Bureau members are supported in IPCC activities by their government or the IPCC trust fund and have professional responsibilities outside of the IPCC, the majority working within research institutes, government departments and/or international organisations. The pressure of time and a lack of financial resources particularly constrain the investment of developing country bureau members (interview 17.09.2010; 20.01.2011). Developing country expertise is highly sought after by international organisations, and these individuals may have to balance IPCC with commitments to other international agencies and assessment efforts (interview 17.09.2010). Historically, the majority of bureau members have not received research assistance to support them in this capacity. As involvement in the IPCC process does not tend to offer developing country participants the same degree of cultural recognition, experts from these regions may be both less able and less willing to invest themselves in the process (Yamineva 2010: 58–59). The economic capital structuring developing country bureau members' capacity to invest in the IPCC process is augmented by the attitudes and perceptions of developed country bureau members, many of whom have historically regarded their counterparts as political appointees (Bolin 2007: 84), not adequately qualified for the task (IAC 2010b, 261 and 587). These judgements overlook the economic resources necessary for a country to (1) become interested and invested in IPCC activities, (2) accrue the cultural capital to meaningfully impact the assessment and (3) have the technical and administrative support to ensure their vision, and representations are incorporated in draft outlines, reports and summaries.

4.4 The Technical Support Units

The construction of each WG assessment report is coordinated and administered by a TSU. Although WG co-chairs are responsible for overseeing the production and approval of the assessment, they would not be able to fulfil this role without considerable organisational, administrative and technical support, as housed within the TSU. As Table 4.5 indicates, the TSUs play a significant role at every stage of the assessments production: preparing and administering the timeline for the report's construction; identifying and processing the selection of authors (Chapter 6); managing the authors in writing the report; editing, harmonising and polishing submitted material (Chapter 6); and compiling the finished product for panel approval and publication (Chapter 7). Although technical support staff are the only unit within the organisation working full time on putting the assessment together, have the most contact with the report as it is assembled, and TSU heads are authoritative figures in and over the IPCC's practice of writing, the TSUs are barely mentioned in the scholarly literature and have not been considered a noteworthy component for analysing and understanding this organisation and the meanings it generates.[13]

The introduction of TSUs to the IPCC's assessment practice is said to have been an innovation of the first WGI chair, Sir John Houghton,[14] when it became apparent that WG specific technical and administrative support would be necessary to realise the finished product (Zillman 2007: 878). These units would subsequently become a central feature of all three WGs and one of the most significant institutional innovations of the IPCC's assessment practice. The TSUs are set up once the decision to repeat the assessment has been taken and the new bureau elected. They are funded by the government of the developed country co-chair and are generally hosted within the chair's institution, such as the university, the met office or the environment agency. To date, there have been eight countries that have hosted TSUs, with both the United Kingdom and the United States holding the WG chair post for five of the six assessment rounds, see Table 4.4.

The WG TSUs are not homogenous units, and although a newly elected chair and appointed staff seek input and advice from outgoing TSUs, the set-up of this unit and the style of work it adopts develop over the course of the assessment as shaped by the WG chair, the appointments they make and the host country. These units have grown over time to keep pace with increasing author numbers and volumes of knowledge, and today they have between five and fifteen members of staff. Nearly all staff will be new hires, as only a few serve on multiple TSU teams and the demands of TSU head make it a difficult role to repeat (interview 14.07.2010). While the TSUs are set up to assist both the developed and developing country co-chairs, this assistance is uneven. The TSU team regularly

[13] For an exception, see Miller 2004; Zillman 2007.
[14] UK chair of WGI for the FAR, SAR and TAR.

update and seek the input of the developing country co-chair, but their main focus is on meeting the requirements of the chair that they work alongside (interview 25.02.2011). The administrative support for developing co-chairs has increased over subsequent assessments, for example, China has co-chaired WGI since the TAR and developed its own technical and administrative capacity within the China Meteorological Administration. However, disparities persist and continue to structure the extent to which a developing country co-chair can invest in the process and imprint on the final product.

The majority of TSU staff are technical and administrative, however, some members of the team are hired specifically for their scientific credentials and experience of previous assessment exercises. The most notable in this regard is the TSU head or the science lead.[15] It is the responsibility of the TSU heads to implement and manage the production of the assessment as envisioned by the WG chair and approved by the panel. The importance of the task is reflected in the credentials of those hired, many of whom are established within a field of science relevant to the WG and have previously contributed as an IPCC author, bureau member and/or a national delegate (interviews 25.07.2010; 5.10.2010).[16] The combined expertise of the WG co-chair and TSU head is critical for gaining the support of the authors. Authors tend to perceive and measure authority through the scientific habitus, if they do not recognise the scientific credentials of those leading the process and trust in their capacity to produce an authoritative assessment, they may invest less in the process. While the scientific capital of the TSU distinguishes it from other administrative units of the organisation, most importantly the secretariat, it is not the unit's main source of capital.

The WG TSUs make an IPCC assessment report possible, binding the assessment practice, and the actors that constitute it, through their day-to-day activities. The TSU's symbolic power lies in the IPCC's dependence on this unit for achieving its mandated task (see Table 4.5). The TSU's position within the organisation gives the unit unrivalled access to the authors and the assessment under construction. The TSU introduces authors to the IPCC and is the main point of contact throughout the assessment. Through emails and author meetings, the TSU staff instil in authors the appropriate procedures and values for conducting the assessment and have the editorial power to ensure that these are adhered to in the compilation of chapters. The TSU's management of the report's construction also

[15] Some TSUs are led by a single head and in other cases the role is split between a scientific/technical lead and an administrative/organisational lead.

[16] For example, Pauline Midgley was head of WGI TSU for the AR5. Pauline has a PhD in atmospheric chemistry and contributed to the science of ozone depletion, publishing articles and participating in international scientific assessments on the effects of CFCs. Prior to her appointment as TSU head, Pauline provided scientific support to the German Federal Ministry of Research, and from 2006 she headed the German IPCC Coordination Office (IPCC-WGI, 2013).

Table 4.5 *The activities and forms of authority of the TSU*

Actor	Activities	Forms of authority
WG TSUs Administrative and technical staff Located in host country (usually the institution of the WG co-chair).	• Support co-chairs and bureaux of respective WG (IPCC 2012a) • Prepare and administer assessment timeline • Process author selection • Manage the authors • Edit, harmonise and polish submitted material • Manage review process • Prepare report, technical summary and SPM for panel acceptance and approval • Finalise for publication	*Economic capital* • Host government *Cultural capital* • **Knowledge of the assessment in process** • Proximity to the assessment, co-chairs and authors • Necessity/centrality to completing assessment • Scientific, technical and administrative expertise *Social capital* • Co-chairs • National focal point and related government office • Relations with secretariat

gives it unmatched technical knowledge of the process and progress of the report, which makes the TSU an important contact point for secretariat, panel and bureau members for informed position taking and decision-making prior to and during bureau and plenary proceedings. This makes establishing and maintaining links to WG TSUs a vital source of social capital and an avenue for acquiring cultural capital (insider knowledge of the process), sources of capital that are most accessible to the member countries hosting these units and the bureau members that work alongside them.

4.5 The Secretariat

The Secretariat is the organisational centre of the IPCC and its only permanent body. Despite its permanence and symbolism as the focal point of the organisation, the secretariat is an enabler rather than a direct contributor to the IPCC's assessment practice. The secretariat plays an active role at the start of the assessment cycle, particularly in assisting the chair and panel in formulating the work program and instilling IPCC values and procedures in the incoming bureau members and TSU staff. However, the secretariats direct involvement in the assessment decreases with the formation of the new WG TSUs. The secretariat is an important actor in plenary and bureau meetings: presenting the agenda and reports of previous sessions, providing support to the chair, introducing budgetary matters, responding to government enquiries and generally ensuring the orderliness of proceedings (see Table 4.6). Between these events the secretariat is regularly in contact with national focal points and bureau members and once the assessment is under way information flows daily between the secretariat and WG TSUs.

Although the secretariat is situated within WMO headquarters in Geneva and its roughly half-dozen staff are employees of the UN, the unit is answerable to member governments of the panel, and it is governments that decide the size and remit of the secretariat (IPCC 2009c).[17] Organisationally, however, the secretariat adheres to WMO procedures in formal administrative and financial arrangements and the secretary reports to the IPCC chair and executive councils of WMO and UNEP (IPCC 2009c). The secretary is an important figurehead within the organisation and to date there have been two long-standing IPCC secretaries, with a third appointed in 2016. These actors have a similar career trajectory to other actors managing the IPCC process, including bureau members, panel members and TSU heads.[18]

In recent years, the authority of the secretariat has been challenged and different factors and events account for this. The distance between the secretariat and the production of IPCC assessment reports has increased with the strengthening of TSUs. As studies of bureaucratic authority indicate, secretariat staff possess a wealth of experience and knowledge, including historical knowledge of the organisation and its policies and procedures. This cultural capital makes the secretariat a valuable contact between plenary and bureau proceedings when government and bureau members seek information and advice from the secretariat to inform decision-making (Bauer 2006; Depledge 2007; Jinnah 2010).[19] While this knowledge is valued within the IPCC, the most valuable form of cultural capital is knowledge of the assessment and its progress in practice, and the secretariat no longer houses science staff and has minimal direct involvement in the day-to-day construction of the assessment reports compared to TSUs. Thus, while the secretariat is the principle point of contact for members of the IPCC and observer

[17] The secretariat was expanded in 2006 (from a staff of five – the same as when it was established), and again in 2009 after a panel review of its staffing and responsibilities. As a result of the IAC review and recommendations the remit and staffing of the secretariat are still under consideration by the panel, see IAC 2010a, 2010d, 8–9; IPCC 2011b.

[18] The first IPCC secretary, Dr Narasimhan Sundararaman, was appointed in advance of the first meeting of the panel in November 1988 and served in this capacity until his retirement in 2002 (Zillman 2007: 877). Dr Sundararaman was a US Federal Aviation Administration Scientist on deputation at WMO and is said to have been one of the "key actors in the decision making process that led to the formation of the IPCC" and influential in the assessment style adopted by the organisation (Agrawala 1998b, 616; interview 17.11.2010). In 2004, Dr Sundararaman was replaced by then deputy secretary, Dr Renate Christ. Prior to her appointment, Dr Renate Christ worked for UNEP, the European Commission and was an Austrian delegate during the development of the Kyoto Protocol. In 2016, Dr Christ was replaced by Abdullah Mokssit, previously the director of the National Meteorological Department of Morocco and national focal point to the IPCC. For a comparison to the career trajectories of the AR5 TSU heads, see footnote 33.

[19] Joanna Depledge (2007) uses the term intellectual capital in her study of the UNFCCC Secretariat and Chairpersons.

organisations, the secretariat cannot provide participants with the same detailed knowledge on the progression of the report as TSU staff.

Between the AR4 (2007) and AR5 (2014), the secretary sought to stem this loss of authority by increasing the scientific capacity of the secretariat and its proximity to the IPCC's assessment practice. However, this brought the secretariat in conflict with TSU staff and led to further erosion of authority. In 2008, the panel set up a task group to undertake a review of the secretariat's staffing requirements, as the unit was widely regarded as over-stretched (IPCC 2008a: 4, 2009c: 2). The secretary, Dr Renate Christ, proposed adding two science officers to the staff and indicated that she saw an expanded role for the secretariat in providing technical and administrative support to the IPCC chair and bureau members on issues and themes that cut across the three working groups and in assessing the grey literature used in reports (IPCC 2009c). The task group dismissed the secretary's request for additional science staff, indicating that:

> ...the working group and task force TSUs are primarily responsible for the preparation of the assessment reports and methodologies and provide the in-house scientific expertise of the IPCC. IPCC interviewees were strongly of the view that the Secretariat should continue to focus on corporate and administrative issues, concerned with the quality and efficiency of processes rather than with their substance. (2009c, 8)

After the panel's review, the post of Scientific Officer in the secretariat was amended to Programme Officer (personal observation). The secretariat's position was further undermined by the media attention surrounding errors over the Himalayan glacier in the AR4 and the resulting InterAcademy Review (IAC), which held the secretariat and IPCC chair responsible for the organisation's 'sluggish response' to these events (IAC 2010a: 47).

In addressing the IAC's recommendations, the panel sought to tighten the remit of the secretariat. This time, however, it was member government's attempts to amend the secretariat's terms of reference that were thwarted. Comments from WMO and UNEP asserted their parental authority over the IPCC, reminding the panel that: 'mutual consent of UNEP and WMO is required to amend the terms of reference of the IPCC Secretariat' (IPCC 2012b 1). Since these events, the secretariat has created a niche for itself in managing external representation of the IPCC and has expanded its expertise in communication and media relations (IPCC n.d.). This extends to providing bureau members and other IPCC actors with training and preparation before media appearances. This demonstrates how units can adapt to changing circumstances to ensure their continued relevance. It also highlights that while scientific expertise and proximity to the assessment are the most valued properties, including within the administration of the organisation, they are not the only activities and forms of authority that matter.

Table 4.6 *The activities and forms of authority of the secretariat*

Actor	Activities	Forms of authority
Secretariat Located in the WMO building in Geneva Between 13–15 members (IPCC n.d.)	• Supports IPCC chair and bureaux (IPCC 2012a) • Manages IPCC Trust Fund • Oversees, organises, and administers plenary meetings, including all documentation • Manages relations between the IPCC and its parent bodies (WMO and UNEP) • Represents IPCC and its products to international stakeholders, most importantly UNFCCC • Manages external communications and media relations	*Economic capital* • Voluntary contributions from Member governments • Contributions from UNEP, WMO, UNFCCC and other international bodies. *Cultural capital* • **Knowledge of IPCC processes and procedures** • Knowledge of (relation with) stakeholders' interests/ investment in the IPCC • Communications and media representation for the organisation *Social Capital* • Member governments • IPCC chair • Bureau • TSUs • Parent organisations: UNEP and WMO • Relation to UNFCCC and other stakeholders

4.6 The Authors

The authors of IPCC assessment reports are largely insulated from the internal social dynamics described above. IPCC authors are experts that have nominated themselves or have been nominated by their government or international organisation and are selected by the WG bureau to assess and review the material relevant to their expertise (Chapter 6) and the government-approved outline (Chapters 5). As with panel and bureau members, producing the assessment is not a full-time job (on paper), and they are not paid by the IPCC for their time. The majority of experts nominated and selected as authors work as knowledge producers and reside within universities, research institutes, government departments and agencies and international governmental and non-governmental organisations. It is from these sites that they contribute to climate change knowledge production, and it is this contribution to a particular body of knowledge, such as the economics of climate change, its health impacts or modelling the general circulation of the atmosphere, which constitute them as climate experts and qualify them to participate in the IPCC's assessment practice. For most authors, their participation in the IPCC is a series of author meetings, email exchanges, and intense periods of reviewing, compiling, assessing and writing to meet the deadlines of the drafting cycle (see Table 4.7).

The social order within the three WGs is largely governed by the scientific habitus, and the forms of authority it recognises, with some variance between WGs depending on the academic composition of its authorship. WGI is the most coherent in epistemic terms. Charged with assessing the physical science of the climate system and climate change, its remit has remained constant since the organisation's establishment. The WG is composed of natural scientists interested in documenting and modelling historic, present and future changes in the composition of the atmosphere, oceans and cryosphere and the relationship to global temperature. The dominance of the natural sciences is reflected in the journal articles referenced in the report, in the TAR, three-quarters of references belonged to Earth science journals, including 'Geosciences', 'Oceanography' and 'Meteorology' (Bjurstrom and Polk 2011: 10). Overall, 84% of references in this report were journal articles, with a small number of journals frequently cited (Bjurstrom and Polk 2011: 4). A similar pattern is likely to be observed in subsequent WGI reports. This highlights that while the focus and coverage of individual reports is shaped by advances in scientific knowledge as scoped and outlined by the co-chairs and approved by the member governments, the production of the assessment remains governed by the shared scientific practices of authors and epistemic conventions for establishing and recognising scientific authority.

The epistemic coherence of WGI is not replicated in the other two WGs. WGII's focus on the impacts of climate change necessitates a multidisciplinary authorship. The majority of WGII authors are again natural scientists, which is reflected in the journal material referenced, the most important fields being the 'Earth sciences', 'Biology' and 'Environmental science' (Bjurstrom and Polk 2011: 10–13). However, WGII covers a broader range of topics and fields of knowledge than WGI, and 'social sciences', 'energy and resources' and 'medicine' are important subjects within the assessment (Bjurstrom and Polk 2011). Furthermore, 59% of WGII references in the TAR are journal articles compared to WG I's 84%, and these references span three times the number of journal titles (Bjurstrom and Polk 2011: 4). This highlights that WGII's assessment of climate change impacts, adaptation and vulnerability relies upon more varied reference material and sources than WGI, including non-peer reviewed material. This is also a reflection of regional chapters and assessment of climate impacts in developing countries, where non-peer reviewed materials are used to fill the gaps in the published literature (interview 7.07.2010). Thus, while the scholarly habitus continues to order relations within the WG and between the chapter team members, the integration of the different fields of knowledge and inequalities in the coverage of knowledge, alongside the negotiation of disciplinary specific scholarly conventions, epistemologies and terminologies are important forces in the conduct of the WGII assessment.

The remit of WGIII has been subject to the most substantial change compared to the other WGs. In the FAR, WGIII was focused on policy response options, and the report was authored by low-level policymakers and negotiators alongside a few independent legal and environmental experts (Skodvin 2000a: 119). In authoring the assessment, this group could not rely on a clearly identifiable body of knowledge for the content of the report or scholarly conventions to structure working relations. As a result, Tora Skodvin suggests that 'the informal rules of politics' became 'natural guides' (Skodvin 2000a: 120). These author meetings effectively served as policy-debating forums – 'where governments could learn about the disputes that would be generated by specific policy options' (Boehmer-Christiansen 1994a: 149), and where preliminary drafting for a convention was undertaken (interview 07.09.2010). After the publication of the FAR, the IPCC's position in the emerging field of climate politics and relation to the international negotiations was not yet formalised (Section 3.1). To ensure the continued relevance of the IPCC's assessment, the leadership at the time adjusted the focus of WGIII to provide an assessment of the cross-cutting economic and other related issues (Bolin 2007: 81). There was also a desire to bring the assessment practices of WGII and WGIII closer to those of WGI and to recruit authors of 'comparable stature' as those within WG I (Bolin in IPCC 1992a: 4), which put economics at the centre of WG III's authorship.

In the end, the bureau's confidence in the political relevance of economics and the scientific authority of economists threatened to undermine the legitimacy of WGIII's contribution to the SAR. Aspects of the economic construction of the climate change proved difficult for countries in the global south to digest, as the statistical value of human life in developing countries was calculated as one tenth of that in developed countries.[20] Nevertheless, economics has remained the dominant form of knowledge in IPCC assessments of climate mitigation (Bjusrstöm and Polk 2011: 11; Corbera et al. 2016; Hughes and Paterson 2017). The social order of relations within WGIII and the conduct of its assessment today are largely governed by the same forms of authority as operate in WGI and WGII: contribution to knowledge (publications), institutional affiliation, and prior IPCC/international assessment experience (Hughes and Paterson 2017). At the same time, the composition of WGIII authorship remains more varied than WGI, including a higher

[20] Chapter 6 of the WGIII SAR used controversial assumptions to calculate the 'social costs' of climate change, suggesting a cash value of $1.5 million to a human life in the OECD against $150,000 in developing countries (Pearce et al. 1996). As a result of developing country objections the final report did not make it through plenary approval and an additional session had to be scheduled (Agrawala 1998b, 626).

percentage of authors from government agencies, international governmental and non-governmental organisations and industry.

The controversy that surrounded WGIII's construction of climate mitigation in the SAR highlights the impact that disciplinary make-up has on the conduct of a WG assessment, the representations of climate change this generates and on the fields of knowledge themselves.[21] Inclusion in IPCC assessment reports demonstrates the social and political relevance of climate change research, and this has made the IPCC an object of competition and struggle within and between different fields of science. Historically, General Circulation Models of the atmosphere have been regarded as the most important scientific and policy tool for knowing and defining the climate change problematic, which is reflected in the number of authors and space given within WGI reports to this form of knowledge (Shackley and Wynne 1995; Shackley et al. 1998; Edwards 1999, 2001; Shackley 1999; Demeritt 2001; Guillemot 2022). Scholarly criticism has identified the limitations of this disciplinary narrowness and the role that social science has to play (Hulme 2008; Nordlund 2008; Yearley 2009; Hulme and Mahoney 2010), although disciplinary diversity has increased, modelling remains central to WGI and WGIII's constructions of future emissions and the climatic and societal responses through Integrated Assessment Models (Beck and Mahony 2018; Cointe 2022). Criticism has also grown over the lack of Indigenous knowledge and representation in IPCC reports (Ford, Vanderbilt and Berrang-Ford 2012; Ford et al. 2016). This is a more challenging issue for the organisation to address as the culture of scientific authority risks further reproducing and entrenching extractive partnerships and practices (Klenk et al. 2017; David-Chavez and Gavin 2018; Latlippe and Klenk 2020; van Bavel, Macdonald and Dorough 2022).

Although the scientific habitus remains an important ordering force within the WG chapter teams; bureau, TSU and panel attempts to increase geographical and gender representation and standardise authorship roles and assessment practices across the three WGs have meant that the authorship has diversified and IPCC's practice of writing has been subject to increasing levels of codification.[22] As covered in depth in Chapter 6, the selection of authors and conduct of early assessments were largely governed by the fields of knowledge and expertise of those that made up the WGs. However, as IPCC reports and scientific findings have been subject to criticism after publication, and as those managing the process have

[21] There has been much interest in the disciplinary compositions of the working groups and the representations of climate change this produces, much of which is critical of the dominance of the physical sciences, see: Cohen et al. 1998; Corbera et al. 2016; Bjüstorm and Polk 2011; Demeritt 2001; Hiramatsu et al. 2008; Hulme and Mahoney 2010; Shackley and Skodvin 1995; Yearley 2009.

[22] For discussion and analysis of this increased codification in terms of STS concerns in formalisation and separation, see Sundqvist et al. 2015.

Table 4.7 *The activities and forms of authority of authors*

Actor	Activities	Forms of Authority
Authors Knowledge producers/ scientific experts on climate change	• Review, assess and compile published knowledge of climate change since last assessment	*Economic capital* • Government and/or institutional support • IPCC trust fund (for developing country authors) *Cultural capital* • **Scientific reputation: contribution to science (publications) + institutional affiliation** • IPCC/international assessment experience *Social capital* • Institutional affiliations • Scientific networks • Bureau • TSU

sought to respond and protect the IPCC and to maintain its authority within the climate field, the scientific habitus has been confronted and at times overruled by other organisational imperatives. To explore this interplay between scientific authority, geographical representation and the codification of the IPCC's practice of writing further, Chapter 6 follows the pathway of the assessment report from the nomination and selection of authors through to the government review.

4.7 Summing Up

In this chapter I have sought to describe the IPCC as I came to understand it through interviews and observation and the analytical framework that developed from this. This is an account of an organisation that has emerged over thirty years and six rounds of assessment and as seen through actor's own description of the everyday activities they undertake in the production of the assessment. Through these activities, actors acquire distinct forms of authority in and over the practice of writing climate change and I describe the valued properties or forms of capital that have emerged to order relations in the IPCC. Historically, relations within the panel were shaped by the epistemic nature of the organisation's mandate and the adherence to scientific conventions and scientific authority by those leading its establishment. However, as member governments have become more familiar and comfortable with their role in the panel and production of an assessment, as the stakes in climate politics have increased and the potential for IPCC knowledge to shape these become more apparent, panel members have asserted their political authority over bureau attempts to contain and channel these forces. While political authority ultimately lies in member governments decision-making power, as with all actors in the IPCC, panel members can accrue IPCC distinct forms of authority through investing time and resources in their specific tasks, such as chairing a task force or contact group and preparing and submitting comments through which a deeper knowledge of the process and social relations with other actors are formed. Following member governments into the production of the assessment over the coming pages enables us to explore how, through participation, delegates acquire symbolic power in the organisation and with what effects in and over the writing of climate change.

Describing the social order within and between the panel, the bureau, the TSUs and the secretariat begins to make clear the value of social relations within the IPCC as a conduit for information sharing and accruing the most valued forms of authority within the organisation: knowledge of the process and proximity to the assessment. It is the organisation's dependence on the co-chairs for realising the assessment, and on the TSUs for its day-to-day production that makes these forms of authority so valuable, after all, the ultimate raison d'être of the IPCC is to produce assessments of climate change. This finding is important to the core concerns of IPCC scholarship. It highlights the value in studying all actors within an organisation rather than privileging those that have recognised forms of power and/or expertise. Thus, while relations within the organisation could not be understood without reference to their scientific and political content, the order of relations and conduct of the organisation are not limited to this. Unique forms of authority emerge within an organisation in response to its mandated task and the necessity of achieving this, as shaped by the actors and field of professional activity engaged to undertake this. Identifying and describing the unique forms of authority in operation within an organisation like the IPCC is also critical for studying the persistence in asymmetry.

As I highlighted from the outset of the chapter, geographical representation is both a core concern to the IPCC and scholars studying this body. Again, it is returning to the historical establishment of the IPCC – those actors that lay the cultural foundations of the organisation – that identifies the properties that were designated of value and which came to organise and order relations within the panel and the bureau. Previous study indicates that while scientific interest and knowledge were growing in the global north, they were not well established in developing countries, where there was both a lack of data and scientific capacity and a focus on other pressing development concerns. This meant that while the IPCC was in formation, many developing country participants were attempting to secure the necessary resources and their continuity to ensure representation at the meetings. This brings to the fore how all forms of authority that are identified and described in this chapter are conditional on having the economic resources to be in the room, to participate and over time to acquire the social relations and forms of authority that meaningful participation is dependent upon. The depth of these asymmetries becomes apparent once we take into account the role of the TSUs, and the extent to which this dedicated technical and administrative unit enables the developed country co-chair to lead at each stage of the assessment's development. However, to really discern this in practice, I need to take you on the journey through the production of an IPCC assessment report, from the outline (Chapter 5), through the order of authorship in the assessment (Chapter 6), to the final approval of its key findings (Chapter 7).

5

A New Assessment Cycle

The aim of this chapter is to begin mapping the pathway that an Intergovernmental Panel on Climate Change (IPCC) assessment report travels as a practice, which means distinguishing the regular activities from those of a particular report or assessment cycle and attempting to weave the story of both. To document a pathway that a report travels in this way is to give your research to a journey of intricate details – scouring document after document to try and establish in whose hands it started, passed through and ended in. When I began reconstructing this journey and following the paper trails archived on the IPCC website, I learned of the multitude of activities that put together the first stage of producing an assessment of climate change: the *outline* of an IPCC report. At first sight, the outline seems like a mundane, even uninteresting element of the IPCC's practice of writing – a list of chapter headings and bullet points identifying the core topics of the next assessment to serve as a guide for the chapter authors (see Table 5.1 for an example). There are four stages to this document's formation: the decision to repeat the process (Section 5.1), the election of the bureau (Section 5.2), the scoping meeting (Section 5.3), and the panel's approval of the final report outline (Section 5.4). Through the unravelling and recounting of each of these stages, however, the web of government and expert input and avenues to influence the content of the next assessment are revealed and the purpose and politics of this list of titles and bullet points come into focus.

In Chapter 4, I described the units of the IPCC and a structure that is fixed, but this chapter captures an organisation in a process of formation. The IPCC reforms with the decision to repeat the assessment. There is continuity in the actors, processes and procedures and the conduct and culture of the organisation, but there is also reflection and re-evaluation at the end of an assessment cycle and change and renewal with the decision to repeat the process, the election of a new bureau and the appointment of technical support units (TSUs). This moment between assessments and the organisational practice for producing the outline allows for those most deeply invested in the IPCC, particularly actors within the bureau and panel, to examine changes in global climate politics and the implications these have on the organisation and for its products. One of the most significant shifts in climate politics took place between the fifth and sixth assessment reports (AR5 and AR6), when a post-Kyoto framework – the Paris Agreement – was negotiated and ratified. The Paris Agreement makes specific mention to IPCC assessment reports as input to the Global Stocktake (GST), as well as the invitation for the special report on the impacts of global warming of 1.5°C (SR1.5) (UNFCCC 2015). The outline's pathway ensures these shifts are captured in the next assessment and by the leadership, assuring the continued relevance of IPCC products (see Table 5.1).

The chapter identifies the central role that member governments have in the scoping and outlining of the next report, from the decision to repeat the process to the election of the bureau, from submitting comments to approval of the final document. Describing the panel's involvement in the production of the outline reveals the avenues that member governments mobilise to influence the election of bureau members and direct the IPCC's next assessment of climate change. This document, however, does not only serve the purposes of member governments; it must also meet the expectations and capture the interests of the scientific community, which will author and validate the report, as well as other stakeholders. This account makes apparent that as with all IPCC documents, the outline serves the purposes and embodies the political and social relations and forces that compile it, which only become visible with intimate knowledge of that practice and the social order shaping it.

5.1 The Decision to Repeat the Process

The practice of writing the outline has developed over time. Many of the features described later were put in place during the scoping and outlining of the SAR and were traversed by each assessment thereafter. In the FAR, the scoping of the assessment was less formalised: terms of reference were established at the first session of the IPCC in November 1988, and these essentially delimited the core topics to be addressed by each WG (IPCC 1988). These terms of reference requested the WG chairs to submit an outline to the bureau within 60–90 days of its establishment (IPCC 1988). In forming an outline, WGI held a scoping meeting that bought together about seventy experts from around the world to agree on chapter headings and outline the contents of the report, the outcome of which was then approved retroactively by the panel (Bolin 2007: 55; interview 1.07.2010). With the completion of the FAR in 1990, and with international negotiations for a framework convention on climate change underway, the continuation of the IPCC, its structure and future work programme became a matter of concern to the institution and its parent bodies.

Table 5.1 *Left: Sample of WGI's outline for the AR5 (IPCC, 2009a); Right: Sample of WGI's outline for the AR6 (IPCC, 2017a). The italics identify reference to the assessment's relevance for the GST; similar references can be found in chapter 1 of the WGII and WGIII outlines for the AR6 (see IPCC, 2017a)*

AR5 WGI Outline *Approved October 2009 (IPCC 2009a)*	**AR6 WGI Outline** *Approved September 2017 (IPCC 2017a)*
Chapter 1: Introduction Executive Summary • Rationale and key concepts of the WG1 contribution • Treatment of uncertainty • Climate change projections since FAR Frequently Asked Questions	**Chapter 1: Framing, Context, Methods** Executive Summary • Synthesis of key findings from AR5 and earlier assessment reports, and connections to AR6 Special Reports • *Framing of the physical science information relevant for mitigation, adaptation, and risk assessment in the context of the Global Stocktake* • Assessment approach • Observational and reanalysis developments since the AR5 • Model and experimental design developments since the AR5 • Emissions and forcing scenarios • Treatment and evaluation of uncertainty throughout the report Frequently Asked Questions
Chapter 2: Observations: Atmosphere and Surface Executive Summary • Changes in surface temperature and soil temperature • Changes in temperature, humidity and clouds • Changes in atmospheric composition • Changes in radiation fields and energy budget • Changes in hydrology, runoff, precipitation and drought • Changes in atmospheric circulation, including wind • Spatial and temporal patterns of climate variability • Changes in extreme events, including tropical and extratropical storms Frequently Asked Questions	**Chapter 2: Changing state of the climate system** Executive Summary • Multi-millennial context, pre-industrial to present day • Natural and anthropogenic forcings • Radiative forcing • Large-scale indicators of observed change in the atmosphere, ocean, cryosphere, land, and biosphere • Modes of variability Frequently Asked Questions
Chapter 3: Observations: Ocean Executive Summary • Changes in ocean temperature and heat content • Ocean salinity change and freshwater fluxes • Sea level change, ocean waves and storm surges • Ocean biogeochemical changes, including ocean acidification • Changes in ocean surface processes • Changes in ocean circulation • Spatial and temporal patterns of ocean variability Frequently Asked Questions	**Chapter 3: Human influence on the climate system** Executive Summary • Overview of model performance and development since the AR5 • Simulated large-scale indicators of change in the atmosphere, ocean, cryosphere, land, and biosphere • Simulated modes of variability • Natural variability versus anthropogenically forced change • Attribution of large-scale observed changes Frequently Asked Questions

The IPCC's FAR proved influential in providing a common scientific understanding of the climate issue and would serve as the basis for negotiations towards a framework convention on climate change (UNGA resolution 45/212 1990).[1] The establishment of the INC, under the auspices of the United Nations General Assembly (UNGA), transferred the responsibility for formulating policy response options from the IPCC's WGIII to this newly formed body, with the IPCC tasked with providing necessary scientific and technical advice to the negotiating process (UNGA res 45/212 1990). The IPCC's relationship to the climate convention was reflected in the WMO's reformulation of the organisation's terms of reference, which charged the IPCC to undertake 'scientific and technical work in support of the negotiations of a framework convention on climate change' and to periodically update 'the assessments of the available scientific information on climate change and the resulting environmental and socio-economic impacts' (Resolution 11 of the WMO congress 1991, see IPCC 2006a, 2007c). In light of these new terms of reference and to help insulate the assessment process from the political environment in which it became situated, the IPCC's practice of writing was subject to codification, and at the fifth session of the panel in 1991, the principles governing IPCC work were formulated (IPCC 1991: 9–9). At this session, a pattern for devising the IPCC's future work programme emerged, laying the foundations of the pathway detailed in this chapter (IPCC 1991).

The assessment pathway formally begins with a panel decision. As an assessment cycle nears completion, the future work programme becomes an item on the panel's agenda and member governments take a formal decision to repeat the assessment process. The documents informing this decision depend on the assessment cycle and whether there is an IPCC chair to guide the process or elections are required, as described in the following section. Either way, the chair's vision paper is one of the first documents produced in the practice of writing climate change. With the support of the secretariat, the chair composes a vision paper on the future work programme and organisational structure of the IPCC to inform government submissions and panel discussions on the IPCC's future work (IPCC 2001a, 2001b; Pachauri 2008). The construction of this vision paper has its own, informal pathway. The product represents the chair's view on the future of the IPCC as informed by bureau discussions, panel members, authors, representatives of the UNFCCC and other international organisations, as well as reflecting commentary taking place in the scientific community (Moss 2000; IPCC 2001a, 2017a, 2017b; Pachauri 2008).[2] While the vision paper centres on the work programme,

[1] See Bolin 2007, chapter 6.

[2] Pachauri (2008: 4) references the discussions in the 'scientific and professional community' on the scale, scope and timeliness of the IPCC assessment process and the suggestions put forward 'which seem to favour a set of focused special reports rather than a comprehensive assessment of the type that has been produced in the past'.

woven within this document are proposals on the appropriate structure of the WGs, the timeline and the content and themes of the next report (IPCC 2001a, 2001b, 2008b, 2008c; Pachauri 2008).

In the case of AR5, the chair's vision highlights the economic and sustainable development aspects of climate change (Pachauri 2008). For the AR6, a new chair – Hoesung Lee – was elected, which meant that the chair's vision paper was circulated after the panel's formal decision to repeat the assessment. When circulated in 2017, the chair's vision highlights the need to shift towards 'applied' and 'solution-focused' assessments that support the implementation of the Paris Agreement and Sustainable Development Goals (SDGs) (IPCC 2017b, annex II). The paper discusses the GST at length, including the timetable for finalising the AR6 in time to inform this newly established process for assessing collective progress towards the Paris Agreement in 2023 and aligning subsequent IPCC assessment cycles to the five yearly cycle of future GSTs (IPCC 2017b: 18–19). This highlights that even before a new bureau is elected or in some cycles, a formal decision is taken, the purpose and content of the next report are taking shape.

The formal decision to repeat the assessment cycle is taken by the panel at plenary session.[3] The panel generally meets annually or biannually in plenary. The sessions are organised by the secretariat, chaired by the IPCC chair and are open to all member governments. They are attended by the bureau, TSU staff, representatives of the parent organisations, WMO and UNEP, and the UNFCCC and other organisations with observer status. Plenary sessions are an important constituent of the IPCC's practice of writing. This coming together at one venue for a three-to-five-day meeting is essential in the formation of a common IPCC identity and shared culture between the distinct units of the organisation. As Chapter 4 describes, it is through routine plenary activities that a collective way of thinking about and conducting the organisation, its assessment practice and an actor's relation to and position on this have emerged.

Plenary sessions take place at different venues around the world by invitation of member governments and are generally hosted in large conference halls where participants are seated in long rows behind alphabetically arranged country

[3] As highlighted above the exact details depend on the chair and the assessment round. For AR4 there were separate plenary meetings for the discussion of the vision paper and the decision to repeat the assessment process (IPCC 2001a, 2001b, 2001c, 2001d). For the AR5 the discussion and decision were held and taken at the same plenary (IPCC 2008a). In the case of the AR6 the decision to repeat the assessment was taken by the panel in February 2015, and the Chair's vision paper and the response by governments and international organisations were presented at the AR6 scoping meeting in May 2017 once a new chair had been elected (IPCC 2017b: 1–2). At the same time as the AR6 was being scoped, a new government task group was established to assess the consequences of the five-yearly GST under the Paris Agreement for the structure and timing of future work, which meant the AR7 was discussed much earlier than in previous cycles (IPCC 2018i).

Figure 5.1 View of the room at the 59th Plenary of the IPCC, Nairobi, 25–28th July 2023. Photo by IISD/ENB | Anastasia Rodopoulou: http://enb.iisd.org/media/ipcc-chair-hoesung-lee-welcomes-delegates-ipcc-59-ipcc59-25jul2023-photo.

plaques or flags, with the IPCC chair and secretariat sat on a podium in front, see Figure 5.1. Each place is provided with a microphone and an earpiece for simultaneous translations into the six UN languages. The meeting schedule is divided between morning, afternoon and evening sessions, and as well as these formal sessions, there is a less formal culture of doing IPCC business in coffee breaks, over lunch and at dinner. These sites of interaction enable delegates, bureau members, secretariat and TSU staff to discuss panel matters and share and shape opinions at a personal level.

The plenary sessions are opened by the chair, who hands the floor to the hosting government and representatives of WMO, UNEP and the UNFCCC (IPCC 2008a). As Neumann (2007) observes in his analysis of ministry speeches, the content of the speech remains largely uniform from plenary to plenary, a practice that enables the speaker to reiterate an organisation's interests in and support for the IPCC and to instil a vision for the forthcoming report. After these speeches, the agenda is approved and the session gets underway. The chair's vision paper and submitted commentary tend to form the basis of discussions on the IPCC's future work programme, and governments raise their country flag to intervene and state their views – or re-state their submitted views – on the subject. Decisions are not usually reached in the full plenary in this manner, instead discussions are moved to contact groups to formulate proposals on key issues, such as the structure of the working groups and organisation and timing of the next assessment. These proposals are then referred back to the plenary for agreement and decision. Contact groups are co-chaired by a developed and developing country member of the panel, as assigned by the chair, and are open to all member governments, although one party delegations are unable to attend parallel sessions.[4] Large contact groups dealing with issues of interest to the majority of the panel are scheduled during the main plenary sessions, where there is translation into all UN languages, otherwise the contact groups proceed in English.

The panel's decision to repeat the assessment process opens the assessment pathway to the next stage in the assembly process: the election of the bureau. This step, the focus of the following section, introduces a new management team to the IPCC's practice of writing, putting in place the necessary professional personnel and administrative machinery required for the production of a global assessment on climate change. The fact that the outline of the report is yet to formally appear on the panel's agenda does not mean that its formation is not underway. When the chair produces a vision paper, and governments, past authors and relevant organisations submit comments, and when these comments are compiled and synthesised by the secretariat informing plenary discussion, the direction and content of the next report is taking shape, orientated by each of these activities and imprinted by the issues, topics and framings that actors write through these tasks. Thus, by the time a new bureau is elected and the scoping of the report formally begins, there are already signposts demarcating preferred directions for the IPCC's next assessment of climate change and criteria identified for those best qualified to lead the process.

5.2 Electing the Bureau

Exploring the outcome of the bureau elections on the distribution of social and cultural forms of capital helps to illuminate the significance of this event and the excitement it generates. While the majority of bureau members are seen as independent from government,[5] bureau membership is an advantage because it enables a country delegate to attend bureau meetings. This increases a member governments social capital, providing increased access to and interaction with the secretariat, chair of the IPCC and those leading and overseeing the next assessment in the WG bureaux and TSUs. These smaller, more collegial proceedings also enable governments to form closer relations with other panel members and offer

[4] Wherever possible this effect is minimised and one-party delegations can request these groups be held separately, although ultimately this is determined by the practical demands of the agenda.

[5] Some bureau members are regarded as political appointees, with 'political instructions from their respective governments' (Bolin 2007: 84).

the opportunity to rehearse decisions and shape their presentation to the panel as well as generate the necessary support for their approval (interview 26.07.2010).[6] These governments accumulate the most valued forms of cultural capital through this testing and formulation of bureau advice and substantive knowledge of the assessment report in progress (see Table 4.1). The additional opportunities to accumulate social and cultural capital that bureau membership offers translate into symbolic power during plenary proceedings and the approval of IPCC documentation through informed interventions that need to be noted and addressed.[7] Bureau membership offers further expertise and insider perspectives during the plenary itself, as bureau members sit alongside and may even speak for the member government during proceedings (interview 4.08.2010).

Holding the most respected positions within the bureau and leading the WG assessments, the election of the developed country WG co-chairs is an important event for the distribution of symbolic power during the assessment cycle. Developed country governments with an elected WG co-chair fund and host the TSU, which, as Section 4.4 described, has greater day-to-day contact with and knowledge of the assessment than any other unit of the IPCC. The office of the national focal point is in regular contact with TSU staff over budgetary and administrative issues, and delegates are likely to seek information, advice and the WG position on plenary issues prior to and during bureau and plenary sessions (interview 20.01.2011). To date, eight countries have hosted a TSU – an investment that enriches them with the most valuable forms of social and cultural capital during the assessment cycle and lasting symbolic power in panel relations (see Table 4.2).

The bureau election also impacts the institutions that support bureau members and the fields of knowledge and professional expertise that authorise them to hold this position. Each WG report is overseen by a WG bureau (see Figure 5.3 for illustration), and the expertise of the co-chairs and six vice-chairs that make up the three WG bureaux, along with the TSU, orientates the direction of the next assessment, delineating the forms of knowledge and epistemic networks accessed in scoping the outline, selecting authors and literature assessed (interview 5.08.2010). This symbolic power to influence is not evenly distributed between bureau members and is governed by bureau position, scientific and/or professional credentials and an actor's investment in the process. Once again, the most significant figure in this regard is the developed country co-chair, as they have the most access to and authority over the emerging assessment and technical and administrative support to implement their vision. Vice-chairs also play a role in scoping the next assessment and identifying regional expertise and colleagues within their epistemic

[6] Bureau meetings are attended by around 50 actors, compared to the 280 or more that attend plenary sessions.
[7] As illustrated in Table 4.2, the six countries that intervened most during the plenary had bureau members.

networks to participate (interviews 4.08.2010; 13.12.2010). However, the extent of their role in decision making and contact with the emerging report depends on individual investment in the process as enabled and encouraged by the co-chairs and TSU and conditioned and shaped by their national institutional setting and the time and resources this provides to invest.

The bureau's impact on the distribution of capital within the panel and the direction of the next assessment means that the elections create considerable activity and excitement, as governments nominate bureau members and lobby for their election. In the case of the most symbolic role in the organisation, the IPCC chair, these election campaigns are visible across social media (see Figure 5.2). The nominating governments promote their candidates through tweets, videos and other social media tools, which are often revealing of the global campaign trail. Until recently however, there was little evidence of the behind-the-scenes manoeuvring that accompanied this element of the IPCC's practice of writing, although the controversy surrounding the election of Rajendra Pachauri in 2002 indicated its extent (see Section 4.2). This changed when Wikileaks provided clear evidence of the importance that some governments place on bureau elections.

Panel and bureau members have attempted to contain the disorder this engenders by codifying election procedures (IPCC 2006b). These stipulate that once the session is open, proceedings pass to a nominations committee, who compile and present candidates to the panel (IPCC 2006b). There has been a long-standing aim to fill bureau positions by consensus rather than taking a formal vote (interview 9.11.2010). Starting with the IPCC chair, followed by the positions of WG co-chairs and vice-chairs, the relevant WMO regional groups meet and attempt to broker agreement on the candidates nominated (IPCC 2006b).[8] Until 2002, this outcome was achieved through the leadership of the chair and the malleability of IPCC organisational structures, which enabled the political interests and geographical representation of the panel to be met through bureau expansion (Bolin 2007: 82–83, 146). However, as interest and investment in climate change and the IPCC have increased, elections have become increasingly reliant on the ballot, and while standing bureau members may want to maintain the spirit of accommodation, behind the scenes governments mobilise all available avenues to influence the outcome.

Research and WikiLeaks on the AR5 bureau election offer us a window on the behind-the-scenes manoeuvring. Observations and interviews by Yulia Yamineva

[8] The overall regional distribution in bureau membership for the AR6 and AR7 is as follows: Region I (Africa) – 7 positions; Region II (Asia) – 6 positions; Region III (South America) – 4 positions; Region IV (North America, Central America and the Caribbean) – 4 members; Region V (South-West Pacific) – 4 positions and Region VI (Europe) – 8 positions (IPCC 2023b).

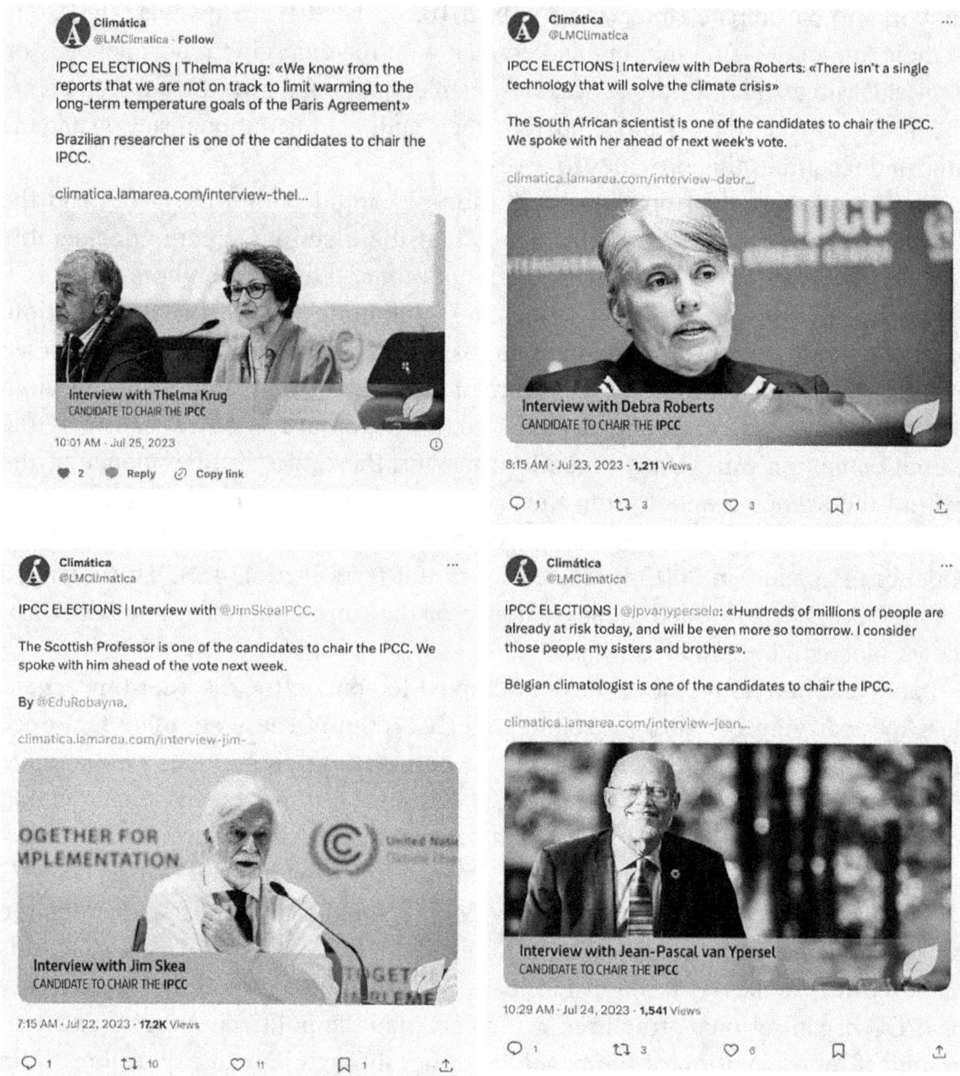

Figure 5.2 Tweets depicting the four candidates in the election campaign for IPCC Chair for the AR7 in 2023. From top left: Thelma Krug (Brazil); Debra Roberts (South Africa); Jim Skea (United Kingdom); Jean-Pascal van Ypersele (Belgium). Tweets by Climática @LMClimatica, 22.07.2023: https://twitter.com/LMClimatica/status/1682635437926481920.

suggested that many delegations arrived at plenary with 'guidance from their ministries of foreign affairs on what countries' candidates to support' (Yamineva 2010: 85), and the WikiLeaks reveal the extent of lobbying by the United States. Embassy cables document US efforts to ensure that their candidate for WGII co-chair (Chris Fields) was elected, but not alongside the proposed Iranian candidate (Mostafa Jafari).

Co-chairing WGII with Iran was perceived as 'problematic and potentially at odds with overall U.S. policy towards Iran', which could 'complicate the U.S. commitment to funding the Working Group Two secretariat' (Guardian 2010a). The United States would not consider withdrawing their nominee for the message it would send to Iran and because 'having a U.S. co-chair at the IPCC significantly bolsters U.S. interests on climate change, a key foreign policy issue' (Guardian 2010a). As this cable indicates, to achieve the desired outcome, the United States sought the support of the IPCC chair and other delegations prior to the election proceedings:

> Prior to arrival in Geneva, *USDEL*[9] *contacted IPCC Chairman Dr.* Rajendra *Pachauri (please protect), who agreed to work on this issue to avoid the potential for disruption to one of the organization's three core working groups* ... Next, USDEL contacted the Austrian delegate serving as EU representative on the nominating committee that manages the election process, who showed an understanding of U.S. equities. USDEL contacted the Malian and Argentinean delegations, who have nominated highly-qualified co-chair candidates (see below), and the German delegation, who have been interested in advancing the Malian for co-chair of Working Group Three, for which Germany has nominated an unopposed candidate as developed-country co-chair.... Also prior to arrival in Geneva, USDEL contacted the UK and Netherlands delegations, both of which we have worked closely with in the past. (Guardian 2010a, italics in original)

In return, the US delegation gave assurances to the countries contacted that it would consider their election outcome preferences (Guardian 2010b, 2010c). This proved effective, with Chris Fields elected opposite the Argentinian candidate, Vincente Barros (see Figure 5.3).

Although political manoeuvring shaped the AR5 bureau election, political interests are not the only force structuring how the IPCC and its assessment practice unfold; maintaining the order of proceedings and the malleability of organisational arrangements continue to act as determinants of eventual outcomes. For instance, in the election of the developed country co-chair for WGI, three candidates were nominated. As a precedence, pressure was applied to candidates and nominating countries to consider withdrawing to avoid a formal vote, as 'it was speculated that a lack of strong consensus for one candidate could potentially be divisive to the work of the IPCC' (IAC 2010b: 245). Nominating delegations did not respond well to this pressure and in the end, candidates were permitted to present themselves to the panel and a formal vote followed, which saw Thomas Stocker of Switzerland elected (Barnett 2008; see Figure 5.3). The adaptability of the organisational structure and proceedings are also apparent in the election of the developing country co-chair for WGIII. Here, rules did not stipulate how to resolve an election result that saw both candidates on the same ballot paper separated by a single vote, an

[9] US Delegation.

Chairman Mr. Rajendra Pachauri (India)

IPCC Vice-chairs		
Mr Ogunlade Davidson (Sierra Leone)	Mr. Jean-Pascal van Ypersele (Belgium)	Mr. Hoesung Lee (Republic of Korea)

WG I The Physical Science Basis	WG II Impacts, Adaptation, Vulnerabilities	WG III Mitigation	Task Force National Greenhouse Gas Inventories
Co-chairs	**Co-chairs**	**Co-chairs**	**Co-chairs**
Mr. Dahe Qin (China)	Mr. Vincente Barros (Argentina)	Mr. Ramon Pichs Madruga (Cuba) Mr. Youba Sokona (Mali)	Ms. Thelma Krug (Brazil)
Mr. Thomas Stocker (Switzerland)	Mr. Christopher Field (USA)	Mr. Ottmar Edenhofer (Germany)	Mr. Taka Hirasishi (Japan)
Vice-chairs	**Vice-chairs**	**Vice-chairs**	**Vice-chairs**
Mr Abdullad Mokssit (Morocco)	Ms. Nirivololona Raholijao (Madagascar)	Mr. Ismail A.R. Elgizouli (Sudan)	
Fatemeh Rahimzadeh (Islamic Republic of Iran)	Mr. Amjad Abdulla (Maldives)	Ms. Suzana Khan Ribeiro (Brazil)	
Mr. Francis Zwiers (Canada)	Mr. Eduardo Calvo Buendia (Peru)	Ms. Antonina I. Boncheva (Mexico)	
Mr. Fredolin T. Tangang (Malaysia)	Mr. Neville Smith (Australia)	Mr. Carlo Carraro (Italy)	
Mr David Wratt (New Zealand)	Mr. Jose M. Moreno (Spain)	Mr. Jim Skea (UK)	
Mr. Jean Jouzel (France)	Mr. Serguei M. Semenov (Russian Federation)		

Figure 5.3 The AR5 IPCC Bureau was elected in September 2008 (IPCC 2008d). As Asia was not represented in WGIII, an additional vice-chair position was subsequently filled by Saudi Arabia.

outcome the panel responded to by enabling both candidates to take up the co-chair position (see Figure 5.3). The resulting AR5 bureau embodied these contrasting forces and highlights that the election of the bureau, like all constituents of the IPCC's practice of writing, is a dynamic interplay between the interests of involved actors (in this case, member governments), IPCC practices and procedures, and the corresponding attitudes and dispositions that investment in the organisation instils.

5.3 Scoping the Next Assessment of Climate Change

With the newly elected bureau in place, momentum for the scoping and outlining of the next assessment report gathers pace, and the pathway widens to make way for the unit that will have greater day-to-day contact with the assessment report than any other unit of the IPCC: the TSUs. Until now, the pathway has concentrated on the activities and decisions taken by the panel at plenary session, and the necessary operations performed by the chair and secretariat to facilitate the decision to repeat the assessment and initiate the process. However, once the bureau has been elected and the TSU assembled, the purpose of each unit becomes more distinct and while their parallel pathways intersect at regular intervals, each unit is focused on its duties. Here, I document the assembly of the TSU before exploring the combined operations of all units at two key events in the outline's production: the scoping meeting and the panel approval of the outline.

It takes about fourteen months from the bureau elections to produce an outline for the next assessment of climate change. In this time, one of the key pieces of machinery required to produce a WG report is put in place. As Chapter 4 indicates, each WG has a TSU, which is funded by the developed country government of the elected WG co-chair and housed in or near their home institution. TSUs are made up of between 5 and 15 members of staff, and it is these units that hold the assessment process together – its timeline, its authors and its contents – to produce an intergovernmentally approvable product. The TSUs sit above and incubate the WG reports from the moment they are assembled until publication, and even when placed into the hands of the authors, the unit maintains a watchful presence over the assessment, editing and polishing the final document. Once the WG co-chairs have been elected, hiring the right staff and assembling this unit becomes the priority of the chairs and those that support them (interview 20.01.2011). The most important hires will be the heads or leads of the unit. Officially, 50% of the chair's time belongs to the assessment, which means that the chair's capacity to fulfil this role rests upon the TSU's ability to manage and conduct the process. To guard the process, the WG co-chair seeks to hire actors with skills, expertise and personal characteristics complementary to their own.

The TSU's first major role in the assessment practice is the scoping meeting. This meeting lasts up to a week and aims to produce a detailed outline of the next assessment report, including the chapter headings and bullet points of the topics to be covered, as in Table 5.1. The meeting centres on identifying and scoping advances in climate change knowledge and in doing so opens the assessment pathway to the scientific communities that will author the report. As such, the meeting also serves as a platform for the newly elected bureau and appointed TSUs to gain the respect and support of those they lead into and rely upon in realising an assessment. For the scoping of the AR5, participants were selected from nominations made by governments and observer organisations and from the expertise identified by the WG bureaux and TSU staff.[10] Out of the 186 experts and government representatives participating, the majority had some prior experience and involvement with the IPCC. As with author selection (Section 6.1), disagreement may arise within the WG bureau during this selection process, particularly over the disciplinary expertise and geographical balance of participants. The scientific and professional expertise of the WG bureau and TSU staff delineate the expert networks accessed and the disciplinary fields represented (interviews 5.08.2010; 20.01.2011). As the developed country co-chair and the TSU have greater contact with and responsibility for identifying expertise and compiling expert lists, the epistemic preferences and geographical range of these personnel can be overrepresented in the participant list unless challenged by the wider bureau (interview 4.08.2010).

The chair's vision paper is the starting document for scoping the next assessment. This document has evolved since its first circulation, reflecting the comments received from member governments, author surveys, plenary discussions and the views of the newly elected bureau (IPCC 2009d, 2017a). Attached to the vision paper is a contribution from each of the WGs that is prepared by the co-chairs and TSUs with input from the wider WG bureau. The WG contributions to the scoping document identify remaining gaps and uncertainties from the last assessment, indicate potential advances in knowledge and include an initial draft outline, or 'straw man' (IPCC 2009d, 2017a).[11] Opened by the IPCC chair, the first day of the scoping meeting is taken up with familiarising participants with IPCC rules and procedures, identifying the main users and target audience of IPCC reports and outlining the initial vision (IPCC 2009e, 2017b, c). The meeting then turns to getting a sense of the state of the field by locating the advances in research, anticipating where further contributions are likely to occur, and finding the means to represent these in the next report (interview 5.10.2010). The ease at which discussion and debate is settled and reflected in the final document depends on the homogeneity of the expert communities and management of the process. It is for instance, easier for WGI participants to identify and agree upon advances in the physical basis of climate change than it is for the diverse range of disciplinary and sectoral expertise constituting WGIII to agree upon the most relevant economic, political, social and ethical dimensions of mitigating climate change (interview 4.08.2010).

[10] Although the InterAcademy review concluded that the scoping process and the selection of participants for the scoping meetings remained 'opaque to those who have not participated' (IAC 2010a: 17), the AR5 scoping process was the most formalised to date. In previous assessments, experts were largely identified and selected by the IPCC chair, the WG co-chairs and TSU staff (IPCC 2003).

[11] The WG TSUs also conduct their own period of consultation with previous authors and relevant experts on the scope of the next assessment report, see for example WGI's background information on the outline (IPCC 2009j, 2017).

The outline generated by the scoping meeting is polished by the co-chairs and TSU staff and sent out to member governments and IPCC observer organisations for comment (IPCC 2009b). Prior to the scheduled approval, the WG bureaux meet to discuss this commentary and revise the outline accordingly. The majority of comments are on the report structure, the timeline and policy relevance (IPCC 2009f). Some comments will identify the use of politically sensitive language, particularly if the topics covered are perceived to relate to the UNFCCC negotiating process (interview 4.08.2010). For the AR5, the WG bureaux took a day to revise the draft outlines, and once presented and approved by the full bureau, they were sent out to member governments along with an information document prepared by the TSU on the context and detail of the outline's production (IPCC 2009g, 2009h, 2009i, 2009j). With the draft in the hands of member governments, focus turns to preparing for the final stage of the outline's formation.

5.4 Approving the Outline

The IPCC process for approving documents is one of the most fascinating facets of the IPCC's practice of writing. Different materials or documents produced by the IPCC are subject to varying levels of 'formal endorsement' by its member governments (IPCC 2013: 2). These start from 'acceptance', which signals that the material presents 'a comprehensive, objective and balanced view of the subject matter', to 'Adoption', where the text is endorsed 'section by section', to the highest level of 'Approval', which 'signifies that the material has been subject to detailed, line by line discussion and agreement' (IPCC 2013: 2). The outline of the next assessment and the final report summary for policymakers (SPM) are subject to the highest level of formal endorsement – line-by-line 'approval' – which, as becomes apparent in Chapter 7, is a process that frequently breaks down into a word-by-word negotiation. In this section, we follow the outline of the next assessment into each of the three WG sessions where it is subject to this process of approval. Although the WG report outlines are much smaller documents in comparison to the SPM, the necessary brevity of chapter titles and bullets "makes each word count just a little bit more" (interview 26.02.2019). As a result, while in some instances there may be mild tweaking, in others, there can be substantial revision by government interventions, objections and suggestions for rearranging the document and rewording the bullet points.

Government's interest and investment in controlling the wording, and thereby potential implications of key findings of an IPCC assessment are central to understanding the struggle that approval sessions generate. The key findings of an assessment report, as presented in the SPM, provide evidence on and warnings about the state of advancement and future projections of climate change, its impacts, adaptation to and mitigation by which methods and on what scale. Once endorsed by member governments through the approval process, these findings become the accepted knowledge base for negotiating the collective response to climate change in the UNFCCC. While the outline only identifies the topics to be covered and is not a widely scrutinised document or an object of the negotiations, its approval can generate the same level of contestation; to understand this, the influence of the final assessment needs to be brought into focus. It is the potential that assessment findings have to impact negotiations – to legitimise or challenge existing objects or to introduce new ones – that governments are sensitive to and seek to guard against (Hughes and Vadrot 2019). The surest way to achieve this is to prevent certain terms, concepts or policies from being assessed in the first place, and this is where the approval of the outline is crucial.

Once approved, the outline constitutes a form of contract: an agreement between the member governments authorising the assessment and the co-chairs leading its production on what the content of the next report will (and will not) cover. As such, the outline can be brought in to play if the final SPM ventures into territory that member governments had sought to avoid by eliminating reference to it in the outline document, as happened in the approval of the SPM for the Special Report on 1.5°C (IPCC 2018a). During the approval, Saudi Arabia expressed 'substantial disagreement' with references in the SR1.5 to Nationally Determined Contributions (NDCs), disagreement they requested was reflected in the report of the session (IPCC 2018b: 6). In making these objections, Saudi Arabia brought the approval of the outline into focus, as recorded in the report of the session:

> The IPCC is providing a scientific basis for governments at all levels. In accordance to the IPCC principles, IPCC products must be policy-relevant and policy neutral and not policy-prescriptive. NDCs and their guidance are currently being negotiated under the United Nations Framework Convention on Climate Change (UNFCCC). Based on this, *the outline* of this Special Report and its scoping were discussed during the deliberations of the Panel on these issues and the Panel agreed *not* to include NDCs in both instances because it would undermine our principles. (IPCC 2018b: 14, italics added)

This statement indicates government's expectation of the approved outline and demonstrates how it has the potential to be deployed later to support objections to and interventions on the SPM text. While instances of this are uncommon, this event highlights the stakes and illuminates the politics in approving the outline.

The WG approval sessions adhere to the same opening routines as all plenaries, although the room may be more crowded than usual, as larger delegations are required to cover the parallel sessions. Once the hosts have been thanked, speeches given and national positions stated, the plenary is suspended and the WG approval sessions begin (Carter, Schulz and Yamineva 2009). These sessions are chaired by the WG co-chairs or a member of the wider bureau, and in general the developed country co-chair leads the proceedings with TSU staff seated next to her or him.[12] The outline is projected on large screens at the front of the hall, and the co-chairs begin the session by identifying the revisions that have been made in response to government comments and suggestions. The session then turns over to the floor, as each chapter heading and bullet point is subject to the scrutiny of delegates, their interventions veering between concern for the political relevancy of the forthcoming assessment to wariness over the political implications of its content. Contact groups are formed to organise the approval and these sessions – focused on particular sections or bullets – are chaired by a developed and developing country member government. Should substantial disagreement over a given chapter or bullet arise, the chair of the session may request dissenting parties to put their heads together in a huddle to agree some acceptable language. Huddles can be formed on the side and/or between proceedings, with the aim of facilitating agreement to be taken back to plenary for approval.

In the case of the AR5, WGIII's outline was subject to substantial revision during the WG approval session in October 2009. WG III is charged with assessing policy options and pathways for mitigating greenhouse gases, and many of the issues arising during the approval centred on the practical utility of the outlined topics and the academic language used to frame the draft (Carter, Schulz and Yamineva 2009: 6–8). Two contact groups were formed, and these groups reordered and reformulated assigned sections of the outline, at times breaking down into smaller groups or huddles, to deal with particularly contentious areas (Carter, Schulz and Yamineva 2009; IPCC 2009k, 2009l). Much of the technical and scientific material assessed by WG III relates to and has implications for negotiations in the UNFCCC. Consequently, many of the tensions that arose were the result of government delegations defending and contesting formulations that could potentially impact the process at a time when a post-Kyoto framework was under negotiation.

In the draft outline, chapter 16 on National and Sub-National Policies separated the analysis of policy implementation and performance into *developed* and *developing* countries, see Table 5.2 (IPCC 2009k, emphasis added). Switzerland raised an objection to these two bullet points, noting that it was difficult to identify a threshold between developed and developing countries (Carter, Schulz and Yamineva 2009: 7). Other countries also intervened, including the Netherlands, the UK and Mexico, suggesting that development

[12] To date, there have been three female WG co-chairs: Valérie Masson-Delmotte (WG I, France) and Debra Roberts (WG II, South Africa) for the AR6 and Susan Solomon (WG I, US) for the AR4.

levels needed to be subject to analysis (Carter, Schulz and Yamineva 2009: 8). Saudi Arabia responded, highlighting that the UNFCCC clearly differentiated between developed and developing countries, as reflected in the division between Annex 1 and non-Annex 1 parties (Carter, Schulz and Yamineva 2009). Such significant categorisations as developed and developing, and the knowledge that they rest upon, have implications for all countries national commitments under the UNFCCC (Dubash, Fleurbaey and Kartha 2014; Edenhofer and Minx 2014). Saudi Arabia did not want its developing country status undermined through IPCC knowledge production processes, particularly at a time when the final report had the potential to impact the negotiations at a key moment in a new agreement's formation. One of the ways to limit this potential is to ensure that these categories are not subject to analysis by authors, an item we revisit in Chapter 7.

While some contestation may be resolved between assessments by decisions taken and agreement reached in the UNFCCC, deep-seated struggles, such as those over developed and developing country differentiation and responsibility for climate action transfers on to new or related objects. This was visible in the approval of the WGIII outline for the AR6 in chapter 15 on Investment and Finance (see Table 5.2). During discussions over the content of the bullets, China and Saudi Arabia suggested adding reference to 'financial flows to developing countries' in the bullet point on investment needs (Mead et al. 2017: 12). This was opposed by the EU, Ireland and Germany on grounds of ensuring political neutrality and to 'avoid being policy prescriptive' (Mead et al. 2017: 12). Ecuador requested to include 'a review of methodologies used to assess financial flows to help ensure objectivity' (Mead et al. 2017). Saudi Arabia and China responded that the UNFCCC and the Paris Agreement mention financial flows to developing countries and emphasised the need to respect this language (Mead et al. 2017). The ENB report of the session demonstrates the bat and ball between these critical elements of negotiation, which requires delegates to be ever attentive to the proposals they offer, how they are received and the discussion and initiatives they invite, which are not always foreseen or easily controlled.

The length of time it takes to approve the outlines depends on the WG and the discussion and debates that surface. The outlines of WGII and WGIII tend to be subject to the most substantial revisions, with sessions running into the night and the approved outline growing in length, as bullets and words are added to capture disparate views and resolve disagreement (see Table 5.2). Although the WGI outline is subject to less revision, this does not mean points of tension do not arise over the direction of the next scientific assessment and its potential to impact UNFCCC negotiations. During the AR5 approval process, China proposed deleting a reference to black carbon in the chapter on clouds and aerosols (Carter,

Table 5.2 *Comparison between the submitted draft outline and the final approved outline for chapters in WG III's contribution to the AR5 and AR6*

AR5 WG III Proposed chapter outline (IPCC 2009k)	AR5 WG III Approved chapter outline (IPCC 2009m)
16. National and Sub-national Policies • Introduction • Taxonomy of policy instruments • Criteria for evaluating policy instruments • Evidence on policy implementation and performance: common experiences across countries • **Evidence on policy implementation and performance: Aspects specific to developed countries** • **Evidence on policy implementation and performance: Aspects specific to developing countries** • Framework: role of institutions and governance • National/state/local linkages • Links to adaptation • Synergies and conflicts among policies • Assessing policy design options	15. National and Sub-national Policies and Institutions • Introduction • Characteristics and classification of policy instruments and packages • Approaches and tools used to evaluate policies and institutions • Research and development policy • **Assessment of the performance of policies and measures in developed and developing countries taking into account development level and capacity** • Framework: Role of institutions and governance • Capacity building • National, state and local linkages • Links to adaptation • Synergies and trade-offs among policies • Assessing policy design options • Investment and finance • The role of public and private sectors and public-private partnership • The role of stakeholders including NGOs • Frequently asked questions

Table 5.2 *(cont.)*

AR6 WG III Proposed chapter outline (IPCC 2017c)	AR6 WG III Approved chapter outline (IPCC 2017a)
Chapter 15: Mobilising finance • Lessons learnt from AR5 and what is new since AR5 • Need for finance – the Paris temperature targets and the NDCs • Public climate finance flows, including multilateral and bilateral, and taking into account effectiveness and scaling up of such flows • International private flows of climate finance • National and sub-national climate finance mobilization and flows, including link to climate policy • Links between national and international finance: Moving the Trillions, including innovative financial mechanisms and public-private partnerships • Successful case studies • The difference in climate-resilient financing consistent with 2, well-below 2 and 1.5 degree scenarios or pathways • Links to adaptation and sustainable development (including co-benefits, synergies and trade-offs) • Financial accountability, including disclosure of climate risks to assets • Emerging trend (e.g. community involvement)	Chapter 15: Investment and finance • Key findings from AR5 and recent developments • Definitions of climate finance • Scenarios of and needs for investment and financial flows related to mitigation pathways and climate change action at the global and regional scales • **Scenarios of and needs for investment and financial flows related to mitigation pathways and climate change action in developing countries** • Investment patterns, and financing for climate resilient development, consistent with different mitigation pathways • Enabling conditions for changing finance and investment patterns • Public climate finance flows, including multilateral and bilateral, taking into account the scaling up of such flows • International private flows of climate finance • Links between national and international finance including developments in financial mechanisms and public-private partnerships • National and sub-national climate finance mobilization and flows, within and across countries, including links to climate policy • Emerging trends (community involvement in climate finance, sustainable investment criteria by institutional investors) • Climate-related investment opportunities and risks • Linkages between finance and investments in adaptation and mitigation, and implications for sustainable development • Case studies

Schulz and Yamineva 2009: 4).[13] This deletion was opposed by the US, UK, Austrian and Canadian delegations, and in the end, China agreed to keep the bullet point unaltered, 'stating that they appreciate the need for an assessment of black carbon but noted that many aerosols also play an important role' (Carter, Schulz and Yamineva 2009).

Struggles over particular elements of the scientific conceptualisation of climate change highlight that these objects can have implications for member governments that are as significant as those that relate directly to mitigation in WGIII. These are struggles over scientific objects that have the potential to become politically weighted through the IPCC's practice of writing climate change. Once the threat of black carbon has been calculated and accepted through the IPCC scientific assessment process, this substance – and the industries that produce it – will be drawn into the political struggle over the international community's response to climate change. Identifying the warming effect that particular gases or particles have on the atmosphere, such as carbon dioxide, methane or black carbon, weights these concepts, and makes scientific terms constituents of global contestation and struggle over emission reduction targets. Those country delegations aware of the political stakes of introducing or highlighting a scientific term in the outline come to the plenary approval session prepared to contest, and if successful, remove these references.[14]

Once the WG outlines have been approved, the plenary is reconvened. The WG co-chairs report back to the panel on their respective approval sessions, highlighting the revisions made and indicating their commitment to the next assessment of climate change (Carter, Schulz and Yamineva 2009: 8). The WG outlines are then accepted by the panel as the outline for the IPCC's next assessment of climate change (with any party disagreement noted in the report of the session), and a timeline for its production agreed.[15]

5.5 Summing Up

This chapter traced the outline's formation from the panel's decision to repeat the process to its acceptance of the final product. The pathway identifies the central role played by member governments at all stages of the outline's development. It is the panel's role in electing the bureau and approving the outline, combined with the IPCC's practice of seeking comments, which enables governments to

[13] The common name for black carbon is soot. These are small light-absorbing particles released into the atmosphere through the incomplete combustion of fossil fuels and biomass. These particles are thought to have both a local cooling effect by reducing the solar radiation that reaches the surface of the earth as well as a regional warming effect through the absorption of sunlight and by the darkening of ice and snow (Forster et al. 2007).

[14] Chapter 7 explores how delegations prepare for approval sessions.

[15] The publication of the WG assessments are staggered to allow the findings from WG I's assessment of the science of climate change to feed into WGs II and III.

imprint at every stage of the outline's formation and shape the topics covered and knowledge surveyed in the next assessment of climate change. This power to influence through the IPCC's practice of writing is not equally distributed among panel members, and an elected bureau member can significantly increase a delegate's access to and authority in and over the process. As such, the bureau election is one of the most politically charged elements of the assessment's assembly pathway. By attending bureau meetings and by having the social and cultural capital of bureau members close at hand during plenary, governments expand their knowledge of the process and capacity for authoritative interventions. This access to the IPCC's practice of writing and knowledge of its proceedings translates into symbolic power during the approval of IPCC documents.

Following the draft outline into the approval session makes apparent why governments seek to maximise their authority in the (re)writing of climate change. As the struggle over chapter 14 and its potential to subject development levels to analysis reveals, how knowledge is assessed and compiled within IPCC reports may have profound implications for elements of negotiation and agreement-making within the UNFCCC. Some forces and tensions may dissipate, as they did between the AR5 and the AR6, once a post-Kyoto framework was reached in the Paris Agreement. However, deep seated contestation – as there is over developed and developing responsibilities and obligations – emerges around new concepts, objects and targets that have the potential to influence the negotiations and here, caution is taken by parties not to re-open for assessment elements that the Paris Agreement settled in their favour.

Once approved, the outline effectively serves as an agreement on the direction of the next assessment between the member governments commissioning the report and the bureau elected to oversee its production. While the outline enables authors to insert their knowledge into the final assessment, it also confines the scope and reach of how climate change will be reported, and any adjustments to approved titles and bullets must be approved by the panel. The assembly pathway as mapped in this chapter, facilitates the creation of a shared vision between all those involved in the assessment's production and serves to harmonise the expectations of the authors with those of the member governments, increasing the likelihood that the panel recognises the outcome and accepts the final product. Now it is time to follow this outline into the hands of the authors at the first lead author meeting, where the bullet points and headings will be transformed into content on climate change. Like all aspects of the IPCC's practice of writing, this pathway through the scientific assessment indicates that while governments aim to structure the reality of climate change through the outline, this reality is subject to re-writing in and through the interests of the authors and other actors that participate in reviewing and redrafting IPCC assessments.

6

The Order of Scientific Assessment

Who has the power to write climate change? As Chapter 5 documented, the writing of climate change begins before the authorship teams are assembled. It is in this chapter, however, that we follow the outline into author meetings, where the content of the next assessment truly takes form and the key findings that will galvanise and orientate future action on climate change emerge. This brings the authors to the fore of the analysis and, importantly, the order of relations that structure the interactions and decisions that imprint on the content of the next report.

Mapping the social order of the scientific assessment is an attempt to understand the social conditions that structure author interactions and the constructions of climate change produced. The chapter aims to help discern who speaks, is heard and leaves a mark on the content of the assessment and what properties authorise some actors to have a greater impact on this writing of climate change than others. Each WG produces an assessment report, a technical summary and an SPM, and this chapter explores the construction of the main WG assessment report. The final report spans thousands of pages and crosses disciplinary divides in its assessment of the science, impacts or mitigation of climate change. Unlike the SPM, which is a widely read and quoted source of information, the WG reports are not widely disseminated beyond the disciplinary and professional fields of authors. As revealed by interviews, the WG reports commonly serve as reference material or teaching aids for scientists and their students, providing a survey of the field and identifying the gaps in knowledge for research agendas and proposals.

The fact that IPCC assessment reports do not serve the same IPCC-specific purposes as the outline and SPM distinguishes them from these documents: the assessment reports are not marked by the political forces governing the panel or the social relations between IPCC units as imprinted on the outline. The reports serve the purpose and embody the relations of the fields of knowledge constructing them and are critical to the transformation of climate change from a scholarly object to an object of political struggle and social action. Mapping the social scientific order of this writing of climate change reveals asymmetry. Following the outline through the panel, it became apparent that the power to write climate change is not equally distributed between member governments. It is in the authorship of the assessment, however, that the inequalities suffused in global climate change knowledge and expertise become apparent. It was interviews and casual conversations that brought this to light. Comments such as only half the author team meaningfully contribute, followed by a pause, as if to say, 'and you know which half', comments that were mirrored in questionnaire responses (IAC 2010b), pointed me towards the exclusions and apparent blindness to it. Although these may have been minority views, these were a powerful minority – actors that historically had a strong voice in the writing of IPCC rules and procedures, the content of the report and in the approval of IPCC documentation. These views are not exceptional to the IPCC. They are common misperceptions that underpin social divisions by nationality, race and gender intersecting with measures of scientific authority. Author relations offer a reflection of the social order of science and broader patterns of global economic and cultural dominance that the IPCC's practice of writing has historically embodied. It is this social order and the misrecognition of the distribution of resources that it rests upon and upholds that this chapter aims to document, alongside organisational attempts to counter it.

6.1 Author Nomination and Selection

Author selection begins with a letter from the secretariat inviting member governments and relevant organisations to 'nominate experts for consideration as Coordinating Lead Authors, Lead Authors, or Review Editors' for the next assessment (IPCC 2010i).[1] Until this point, the IPCC's practice of writing has remained largely closed to the scientific communities that generate climate change knowledge. During the scoping of the report, expert input comes from the WG bureaux and scientists nominated by their government and invited to the scoping meeting, many of whom are known through participation in previous assessments. The secretariat's request for author nominations and the roughly three-month nomination period initiate a wider search for qualified authors, providing the opportunity for national focal points and relevant international organisations to identify experts and for climate change experts to identify themselves and gain access to the IPCC's practice of writing.[2] Not all qualified experts will be aware of this process, be nominated by their government or live in a country that submits author

[1] Before the Fourth Assessment Report (AR4), authors were selected prior to the finalisation of the outline. However, this procedure was altered in part because those managing the TAR discovered that they did not have the appropriate expertise to fulfil sections of the outline (Yamineva 2010: 54–55). See also IPCC 1997.

[2] In earlier assessments self-nomination was 'the norm' (IPCC 2010d: 7).

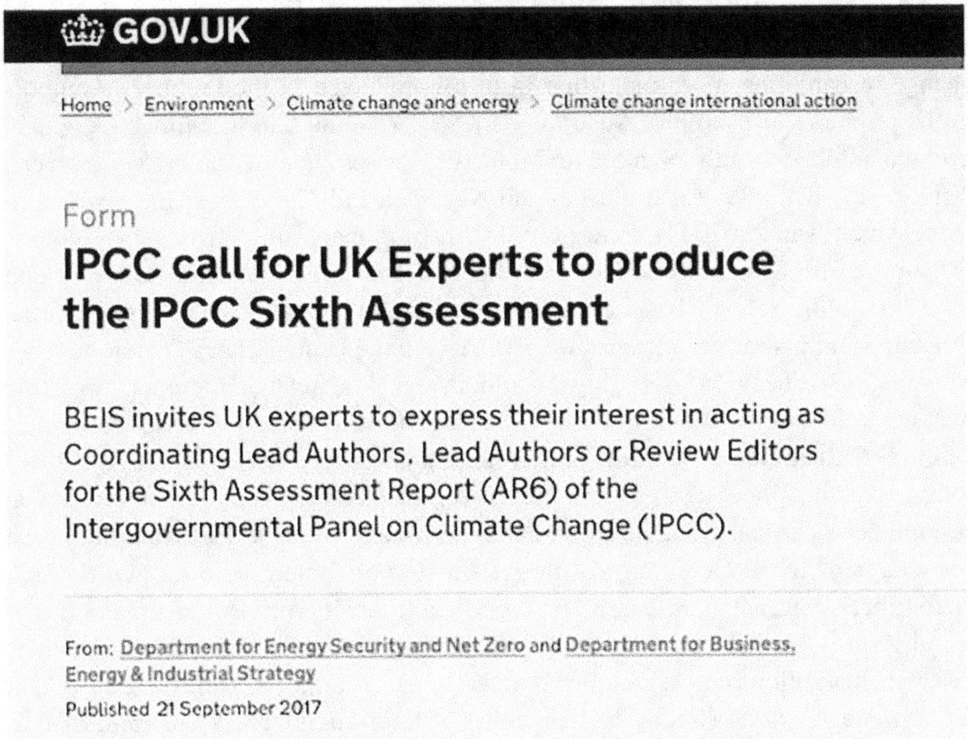

GOV.UK

Home > Environment > Climate change and energy > Climate change international action

Form

IPCC call for UK Experts to produce the IPCC Sixth Assessment

BEIS invites UK experts to express their interest in acting as Coordinating Lead Authors, Lead Authors or Review Editors for the Sixth Assessment Report (AR6) of the Intergovernmental Panel on Climate Change (IPCC).

From: Department for Energy Security and Net Zero and Department for Business, Energy & Industrial Strategy

Published 21 September 2017

Figure 6.1 UK call and application form for experts for the AR6. Screenshot taken from UK government website (From the UK Department for Energy Security and Net Zero and Department for Business, Energy and Industrial Strategy, published 21.09.2017). This contains public sector information licensed under the Open Government Licence v3.0. Available at: www.gov.uk/government/publications/ipcc-call-for-uk-experts-to-produce-the-ipcc-sixth-assessment.

nominations, and at this stage asymmetries in access and participation become apparent.

The focal point is regarded as the link between the government and the national scientific community, and as such it is responsible for orchestrating the national process for identifying experts and submitting a government-approved list of author nominations. How this list is compiled depends on the national context (IPCC 2010c). The IPCC encourages focal points to keep data bases of past authors and reviewers to contact (IPCC 2010c), and many developed country focal points have support staff and well-established mechanisms for sending out the call to government agencies and academic networks (see Figure 6.1), which may extend to workshops to raise awareness of IPCC authorship (IPCC 2010c: 6). Some developed countries have a set of institutionalised procedures, such as government and community consultations or expert committees, which are convened to assist in the selection process (IPCC 2010c: 7). This investment in author selection indicates the recognition that the government gives to the IPCC as a platform for national climate change research, an investment that in turn is reflected in the number of authors in the report.[3] For example, Australia and New Zealand had five times the global average number of authors in the first four IPCC assessment reports (Ho-Lem et al. 2011: 1311–12), with Australia 6th and New Zealand 15th across the six assessment cycles (Tandon 2023). In terms of the institutional affiliation of these authors, the Australian Commonwealth Scientific and Industrial Research Organisation (CSIRO) is the second best represented institution overall, behind the National Oceanic Administrative Organisation (NOAA) in the United States (Tandon 2023). These figures reflect long-standing Australian and New Zealand membership on the bureau, national investment in the IPCC and institutionalised processes for undertaking IPCC activities, including author nomination (interview 25.07.2010).

Existing links between the national government and climate change research communities can narrow the search for expertise. In the UK, for example, in the early days of the IPCC, many of the scientists nominated were known through government-contracted research for the Department of Environment, Food and Rural Affairs. These actors were drawn from or related to a small number of research institutions, most notably the Met Office Hadley Centre in Exeter and the Climate Research Unit at the University of East Anglia, and were connected to each other and the focal point through these locations, academic networks and participation in research projects and national climate change assessments (interviews 30.06.2010; 07.07.2010b; 21.09.2010).[4] Although the call for author nominations is more widely distributed and anticipated in the UK, traditional centres remain better represented in appointed authors (Corbera et al. 2016; Tandon 2023).[5]

In contrast, scientists from developing countries have raised concerns about the responsiveness of their focal point and the appropriateness of the expertise nominated 'either because they do not know who those scientists are or because

[3] For example, the report on UK-funded research on climate change and international development identifies the IPCC and authorship as an important measure of impact (Scott et al. 2021). A memo released from Environment Canada's science and technology branch highlights the significance of Canada's contribution to the IPCC for maintaining the country's reputation as a leader in climate change research (de Souza 2010).

[4] The UK has had an active role in the management and organisation of the IPCC since its founding. Sir David Warrilow, who retired as UK focal point in 2016, attended IPCC plenary's since 1990, and Sir John Houghton, the director-general and later chief executive of the UK Met Office, chaired WGI's contribution to the FAR, SAR and TAR. It is therefore unsurprising that there are well-traversed routes between the UK focal point and the climate change research community, which are re-activated with each subsequent round of an IPCC assessment of climate change. For an account of the relationship between the then Department of the Environment and the Met Office, see Shackley 1999.

[5] For the AR5, the government's call for nominations was answered by 154 applications, which after review by a small panel of experts from DECC and elsewhere were all put forward for the IPCC selection process (see DECC, 2010). Institutions of the 63 authors selected: 8 Met Office Hadley Centre; 7 University of Exeter; 6 Oxford University; 6 Cambridge University; and 5 University of East Anglia (IPCC 2011c). See also Corbera et al. 2016; Tandon 2023.

political considerations are given more weight than scientific qualifications' (IAC 2010a: 18). The lack of a coordinated effort for author nomination and selection in developing countries is supported by an IPCC study, which indicates that of those surveyed only half of developing and EIT country focal points submitted nominations for the AR4, compared to over ninety percent of those from developed countries (IPCC 2009n).[6] The first report by the special committee on developing country participation, published in 1992, indicates that the degree of coordination between various departments and ministries of governments and the 'manpower' [sic] 'to receive, communicate and disseminate information' was often not sufficient within developing countries (IPCC 1992b: 157), which meant that from the outset there was difficulty establishing the processes to fulfil these requests and tasks. As a result, developing country authors have sought alternative avenues, either through nomination by an international organisation or through a developed country focal point (Tandon 2023).

Once the nomination period closes, it is the task of the TSUs and WG bureaux to select and compile the author teams. This selection process has become a significant undertaking. Since the publication of the FAR in 1990, there has been rapid growth in scholarly interest in climate change, which has led to a corresponding increase in the number of experts qualified to author IPCC reports, as well as the volume of literature to be assessed.[7] The three WG reports of the IPCC's FAR totalled 940-pages of climate change assessment. The WGI report was compiled by 35 authors drawn from 12 countries, reflecting the fact that academic interest in climate change was emerging and largely confined to a few research centres in the UK (Boehmer-Christiansen 1995a, 1995b; Shackley 1999) and North America (Edwards and Lahsen 1999).[8] This in turn meant that the available literature on the subject was limited and could effectively be reviewed by one or two leading experts with requests for contributions where necessary (interviews 1.07.2010; 21.09.2010; 26.10.2010).

There were over 3,000 author nominations received for the AR5, an increase of 50% from the previous assessment (IPCC 2010d), and 2,858 nominations from 105

[6] At the 30th session of the IPCC in Antalya (IPCC 2009a), the IPCC vice-chairs were charged with assessing the involvement of developing/EIT country scientists in order to make recommendations for improving participation. As part of the analysis a survey was conducted, which of the 194 IPCC member countries only 38 responded, and of those respondents from developing countries (18) and EIT (4), 50% indicated that no experts were nominated for AR4 authorship (IPCC 2009n). This would suggest that in actuality the figures are even lower, as those responding to the questionnaire are probably more invested in the process.

[7] According to a survey by Boehmer-Christiansen and Skea (1994: 20–22) growth in knowledge was already observable between the FAR and the SAR, particularly in the fields of climate modeling, climatology, oceanography and the physical impacts of climate change. A UKCDS review indicates the scale of this increase over the past two decades: 'between 1992–1996 the total global sample of publications was 2,467, over the period 2007–2011 this has now risen to 27,055' (McLaren and Carter 2010: 15).

[8] Three developing countries were represented with a single Lead Author from Brazil, China and India (see Houghton et al. 1990).

countries for the AR6 (IPCC, n.d.). Through the selection process, the WG TSUs must ensure that the criteria used to assess professional credentials and differentiate between applicants identify the leading scientific authorities from expanding fields of climate research and compiles author teams with the necessary expertise to address each heading and bullet point of the panel-approved outline (interviews 5.10.2010; 20.01.2011). Conventions for distinguishing scientific authority are the first criteria used to identify suitable candidates. It is the job of the TSU to measure an applicant's research impact and productivity. The tools for this vary across the WGs and assessment cycles and include h-index, i10-index, Research Gate scores and number of citations as recorded on Google Scholar (IPCC 2010e, 2018c, 2018d). WG bureaux may have other considerations they want to include in the selection procedure. In the AR5, WGII consulted 'respected scientists and researchers' to evaluate chapter candidates, particularly in regard to young scientists that were new to the IPCC and not known to WG bureau members (IPCC 2010f: 1). For the selection of chapter leaders, the AR6 WGIII bureau considered 'qualities of leadership required to lead multidisciplinary and international chapter teams' alongside scientific excellence (IPCC 2018e). At this stage, the emerging shortlist may need further supplementing by bureau and TSU members to ensure the necessary knowledge and regional representation is captured (IPCC 2010g; interviews 5.10.2010; 20.01.2011). Some authors may be contacted informally to encourage participation, discuss a leadership role and ensure support for the process (interviews 20.11.2010; 20.01.2011).

However, it is not solely WG priorities and scientific measures of authority that govern author selection. The final shortlist must meet the IPCC's criteria for including the range of views, geographical representation, experts with and without previous IPCC experience and gender balance (IPCC 2013: 5–6). These criteria arise from the IPCC's attempt to ensure continued acceptance and support for the assessment process amongst its member governments and the communities of knowledge that evaluate the finished products. Geographical representation has been a central concern to IPCC leadership since its establishment and a number of institutional mechanisms are in place to enhance developing country membership to the panel and in the authorship of the reports.[9] These mechanisms, such as bureau members identifying regional expertise to supplement government

[9] The first chairman of the IPCC, Bert Bolin, is often quoted for his remark that 'right now, many countries, especially developing countries, simply do not trust assessments in which their scientists and policymakers have not participated. Don't you think global credibility demands global representation?' (Schneider 1991). The IPCC convened a Special Committee on Participation of Developing Countries in 1989 that made a number of recommendations, including offering financial support to developing country participants to attend plenary meetings and working group sessions (Lunde 1991: 77–78; IPCC 1992b; Skodvin 2000a: 130–31). Participation of developing countries in IPCC activities remains a regular item on the agenda (IPCC 2009n, 2018h).

nominated applicants, adjustment to measures of expertise to fit the regional contexts (Standring and Lidskog 2021: 11–12), ensuring that there is at least one lead author 'and normally two or more from developing countries' on each chapter (IPCC 2013: 5) and funding travel costs (IPCC 1992b: 152), have contributed to an increase in the selection of developing country authors across all WGs (IPCC 2018c, 2018d, 2018e). Despite these increases, however, significant issues remain in the identification and nomination of developing country authors (interviews 5.10.2010; 20.01.2011; IPCC 2019), which reappear throughout the authorship of climate knowledge.

Each of the selection criteria has a history and is in part a response to scrutiny and criticism. The criteria for balancing the retention of experienced authors with the introduction of fresh insight, for example, came into focus during author selection for AR5. The release of emails from the Climate Research Unit at the University of East Anglia reinforced existing perceptions that the IPCC assessment process was governed by a few long-standing members seeking to keep critical science out of the reports (de Costello 2009; McIntyre 2009; Pearce 2010; Jowitt 2011). The need for criteria on gender balance advanced through several avenues, including general observation and discussion (IPCC 2018f), data on gender disparity in the organisation (Gay-Antaki and Liverman 2018; Gay-Antaki 2021; Liverman et al. 2022) and UN level consideration and targets (IPCC 2018f). The significant increase in the number of women in the assessment from 8% in the FAR to 32% in the AR6 demonstrates how increased awareness and organisational criteria impact author selection (IPCC 2019: 13).

The final shortlists are drawn up over email and via teleconferencing between co-chairs, the wider WG bureau and TSU staff; with the particulars of the process and the extent of bureau involvement dependent on the WG. Contention can arise at this stage as the structural disparities between developed and developing country scientists and measures of scientific credentials surface. All three WGs used publication output and impact factor to measure nominees' scientific authority for the AR5 and AR6. These measures of actors' scientific credentials, along with institutional affiliation, are the same sources of cultural capital that order social relations within the fields of knowledge themselves, as well as in the bureau and chapter teams (see Tables 4.4 and 4.8; Tandon 2021, 2023). Studies of the global economy of knowledge illuminate the structural disparities that scientists from the global south are subject to in acquiring these forms of scientific capital (Paasi 2005; Jöns and Hoyler 2013; Collyer 2016; Connell et al. 2018a, 2018b), which 'assume there is a homogeneous domain of knowledge on which the measuring operations may be performed' (Connell 2014: 211).

Institutional affiliation is often the first criteria visible on a CV. Research identifies the Northern elite university system as central to the global production of knowledge, including in new fields like climate change (Corbera et al. 2016; Connell et al. 2018a). This is also evidenced in the Shanghai and Times Higher Education-QS university rankings (Jöns and Hoyler 2013). Out of the 100 highest ranked institutions in 2009, over half of the Shanghai list (55) and one-third of the THE-QS (32) were located in the United States, and 11 of the Shanghai and 18 of the THE-QS in the UK (Jöns and Hoyler 2013: 50). Moving down the CV from institutional affiliation to publication record as a measure of scientific authority reinforces this pattern. Publishing practices are 'fundamental' to knowledge production (Collyer 2016: 69), and Anglo-American institutional dominance is mirrored in journal rankings and citation data. In a 2003 study of journals listed by the Thomson Institute of Scientific Information (ISI), an index generated by a US-based firm, the United States and the UK in particular, and English-speaking countries overall, dominated the total number of ISI indexed journals (Paasi 2005 779–80), with 66% of the journals in the science database from English-speaking countries, which rose to 85% for the social sciences (Paasi 2005).

The majority of these journals are edited in the United States, which makes publishing in them more accessible to US academics, who are both writing in their mother tongue and submitting to a peer review process that is 'dominated by people accustomed to both the language and methodology of US scholars' (Altbach 2004: 10–11). This puts scientists that are communicating in a second language and conforming to less-familiar academic norms at a significant disadvantage (Altbach 2004; Collyer 2016; Tandon 2021).[10] In interviews, journal editors cite the lack of awareness of current literature and the choice of outdated methodologies as holding back developing country submissions (Altbach 2004: 9). Older sources indicate editorial discrimination against authors on the basis of institutional affiliation and where English is not the first language (Gibbs 1995: 96–98). As a study by Collyer (2016) identifies, these editorial biases persist, with one US male editor of a sociology journal suggesting that 'we very rarely get serious papers from developing countries … it is just a different kind of world'. He extended this judgement globally, suggesting 'there are very different styles of work in different countries, much of which is "not to our taste"' (Collyer 2016: 65). Editorial teams are in some cases taking steps to acknowledge and challenge these asymmetries (Schipper et al. 2021) and transregional circuits of publication have emerged (Collyer 2016). However, these are steps against a tide of commercialisation, with 70% of the top 57 publishing companies headquartered in North America, Europe and the UK (Collyer 2016: 64), resulting in the further standardisation of international editorial and publishing

[10] A 2002 study by UNESCO suggests that developing countries have 17.5% of the world's share of scientific publications (UNESCO 2005 cited in Yamineva 2010: 60). A survey of the journal science in 1995 indicates it only accepted 1.4 % submissions from 12 of the most prolific developing countries, which was the same figure as in 1991, despite a doubling in the rate of submissions (Gibbs 1995: 96).

practices (Collyer 2016). Even when published in international journals, scientists outside of the US and Europe are less frequently cited (Gibbs 1995: 98; Jons and Hayler 2013: 53–54), which impacts a scientist's citation index and perceived contribution to knowledge. It is therefore unsurprising that in Reuter's ranking of the top 1,000 climate scientist, over three quarters were located in Europe and North America, and only five were listed for Africa (Okem et al. 2021), four of which were in South Africa (Reuters 2021).

To understand the persistent asymmetries in the distribution of scientific capital between developed and developing countries – as measured by these indices – the dependent relationship between scientific authority and economic capital has to be brought into focus. Contributing to knowledge through academic research requires access to libraries, databases, laboratories, office space, administrative support, internet and other electronic resources, and this substantial financial and infrastructural support must be consistent and long-term (Altbach 2004). The cost of maintaining a research university continues to grow because of the increasing complexity and cost of scientific research (Altbach 2004: 7). Studies on research expenditure and its link to output reveal how economic resources, as measured by Gross Domestic Product (GDP), impede developing countries from generating their own climate knowledge (Ho-Lem et al. 2011; Pasgaard and Strange 2013). Thus, while the United States spends more than 2.5% of its annual GDP on research and development, 'no country in sub-Saharan Africa – even the comparably rich South Africa – spends more than 1%' (Tandon 2021). This makes external funding critical to building knowledge on climate change for the majority of African countries, the consequence being that national knowledge production is shaped by external interests. Overland et al. (2022: 710) estimate that between 1990 and 2020 USD 620 million was spent on funding Africa-related climate research. The main sources for this were the United States, the United Kingdom and the European Union, with research on the continent largely being 'dictated by the priorities' of government institutions in these countries (Overland et al. 2022).

Expenditure on research is particularly relevant to climate knowledge because computing power has become essential for modelling future climate change and potential response pathways. Participating in and contributing to these forms of knowledge requires substantial investment in computer hardware, processing power, data, programs for producing and running models and high-speed internet to share and download the above. Even in highly industrialised countries with long-term investment in climate science, such as Australia, a researcher remarked that there is 'just one computer system in the whole country advanced enough to handle this work' (Connell et al. 2018b: 10). The dependent relationship between economic and scientific capital provides an important explanation for Northern dominance in the production of climate change knowledge, and the preponderance of scientific authority as measured by publications and impact (Sagar and Kandlikar 1997; Kandlikar and Sagar 1999; Lahsen 2004; Karlsson, Srebotnjak and Gonzales 2007; Pasgaard et al. 2015; Blicharska et al. 2017; Schipper et al. 2021; Overland et al. 2022).

Alongside and in some cases, as a by-product of resource asymmetries, there is the national context in which researchers work that are productive of place-specific academic cultures (Borland, Morrell and Watson 2018; Ibarra et al. 2022). There may be limited recognition for the extra effort required to publish internationally, and instead, incentives to publish in national or regional outlets as well as national measures of contribution to knowledge and career progression (Collyer 2016; Tandon 2023). The pay and conditions, alongside the national funding context, may also mean that academics and institutions need to supplement incomes through consultancy work or because of the competition for limited expertise are involved in other national, regional and international assessment exercises (interview 17.09.2010). The resource asymmetries and the constraints they create need to be viewed alongside conscious decision-making to invest in local and national development priorities with industry and policy partners with place specific rewards over international scientific collaboration and publication (Borland, Morrell and Watson 2018).

It is during bureau scrutiny of the author list that there is opportunity to consider these multiple intervening factors and their effects. Historically, however, the lack of nominations for authors from developing countries, alongside scientific criteria for selecting nominees has been compounded by dispositions within the bureau. For instance, when interviewed some bureau members suggested that their counterparts in developing countries were not forceful or articulate enough when it came to contesting items on the agenda (interview 13.12.2010). In some cases, this was identified as part of a wider malaise in the bureau, characterised by a lack of interest and contribution from developing country bureau members (interviews 9.11.2010; 13.12.2010), which was sometimes put down to political appointments.[11] Comments like these and those critical of the overly formal style of developing country participants are indicative of the culture that has historically ordered relations within the panel and bureau as described in Chapter 4, which overlooks the resource maldistribution that this culture is built upon and serves to maintain. This order of relations is taken as a given and those not conducting themselves accordingly or presenting their grievances appropriately have historically

[11] Such views are also prevalent in responses to the IAC questionnaire, for example one bureau member suggests that the bureau is ‘too geopolitical’ and goes on to say: ‘I tried very hard to engage my WG2 bureau in author screening/selection, critical review of the zero order drafts, etc., and only one out of six were really helpful. Two others meant well, but didn’t know the science well enough to be constructive, and the other three were simply unprepared to help in any meaningful way’ (IAC 2010b: 587). Another IPCC participant with an insider view of the bureau suggests that ‘too much consideration of regional balance and balance between developed and developing countries spoils academic integrity, in some cases’ (IAC 2010b: 261).

struggled to have their contestations and contributions acknowledged within the IPCC's practice of writing.

The final lists of WG authors are subject to full bureau and plenary approval. During these proceedings grievances over selection criteria and regional representation can again resurface, and if the bureau decides there is insufficient increase in the number of developing country authors from the previous assessment, WGs may be requested to reconsider the geographical balance of authors and amend accordingly, as WGI were requested for AR5 author selection (IPCC 2010h). At this point, those that have spent time compiling and amending these lists express frustration, highlighting the impossibility of such a task, and bringing attention to the number of developing country focal points that did not submit expert nominations (interview 20.01.2011). It is only once author lists have been revised and accepted by the bureau, however, that the assessment progresses. With bureau approval, author lists are made public and appointment letters are sent to successful candidates. Those that are unsuccessful are added to the IPCC database and requested to review the emerging report later in its assembly.

6.2 The Authorship

There is a hierarchical structure to the authorship of IPCC reports, and the aim of the following section is to detail the historical development and breakdown of labour between the coordinating lead authors (CLAs), lead authors (LAs), contributing authors (CAs), and review editors (REs). The articulation of these titles and the duties they entail have resulted from bureau and panel attempts to learn from the experience of previous assessments and in response to criticism sustained after the report's publication (interview 5.08.2010). The IPCC's assessment practice has been subject to scrutiny by those seeking to undermine its conclusions. These criticisms were particularly vociferous after the publication of the SAR in 1995, and in 2009, when emails between IPCC authors were made public and errors were discovered in the regional chapters of WGII's contribution to the AR4. To respond to these events and prevent similar incidents undermining the next assessment, authorship roles and rules of procedure have been periodically updated and codified.[12]

The FAR and SAR were put together and overseen by fewer participants and management of this process, such as that required for compiling the author teams, was largely the responsibility of the developed country chair and the TSU, with assistance and advice from key members of the then smaller WG bureau. Within

12 The tasks and responsibilities of authors were clarified after the FAR (IPCC 1993: Appendix G, Annex 2). Then after the SAR, IPCC procedures for the preparation, review, acceptance, adoption, approval and publication of IPCC reports were adopted at the Fifteenth Session (IPCC 1999), and have been updated and amended regularly since then, see IPCC 2013.

the author teams, roles were not formally assigned, and leadership of the chapters was established more through scientific authority than formal decision-making (interview 26.07.2010). In addition, actors requested by members of the chapter team to contribute material were oftentimes colleagues within the same research institution and could become formal members of the chapter team through these informal avenues (interview 26.07.2010). In the first two assessments then, the scientific culture of authors governed the process (interview 1.07.2010, 21.12.2010). However, as climate change has ascended the political agenda, pressure on the IPCC and its conclusions increased and governments became more involved in the work and leadership of the panel (interview 13.12.2010). As a result, the informal epistemic conventions governing the IPCC's assessment practice have been subject to member government scrutiny and codification.

The SAR's conclusions were undermined by criticism surrounding IPCC procedures for revising the WGI report (Lahsen 1998; Skodvin 2000b; Edwards and Schneider 2001; Houghton 2008). The controversy was initiated by an op-ed piece in the *Wall Street Journal*, which followed a similar criticism made earlier by the Global Climate Coalition (*Nature* 13 June 1996: 539). In this piece, an American physicist accused WGI lead authors of seriously corrupting the peer-review process by altering the text of the assessment after it had been formally 'accepted' by the panel (Seitz 1996). The chapter in question was the source of a widely quoted sentence from WGI's SPM, which stated that despite large remaining uncertainties, 'the balance of evidence suggests that there is a discernible human influence on global climate' (Houghton et al. 1996: 5). The debate lasted several months, turning into a disagreement over the underlying scientific evidence for the statement. The exchanges between IPCC officials and their critics revealed ambiguity in the IPCC rules of procedure, which neither allowed nor prohibited changes to a report after its formal acceptance (Edwards and Schneider 2001: 227).

In another incident in the SAR, WGIII's report got stuck in approval proceedings due to objections from the developing countries to controversial economic assumptions used in the calculation of the 'social costs' of climate change (Agrawala 1998b: 626). In the chapter, a cash value of $1.5 million was assigned to a human life in OECD countries against $150,000 in developing countries (Pearce et al. 1996). These controversial calculations could have been avoided if there had been better oversight of the authors' response to review comments (Agrawala 1998b: 626). In 1999 and in preparation for the third assessment report (TAR), the IPCC tightened its rules of procedures surrounding the approval and amending of text and introduced review editors to the authorship of IPCC assessments (Agrawala 1998b: 228–29; Skodvin 2000b).[13] These procedures and the role of

[13] The role of review editor was previously undertaken by the WG vice-chairs and TSUs.

review editors were subject to further scrutiny after the publication of AR4 in 2009, when mistakes were discovered in WGII's contribution concerning the melting rate of Himalayan glaciers and the land area below sea level in the Netherlands (IAC Review 2010a; PBL 2010). The panel again responded to these events and the surrounding international media criticism by tightening the assessment rules and procedures (IPCC 2010a, 2011a).

Today, and as a result of these events, the responsibilities and duties of IPCC authors are codified in IPCC rules and procedures (IPCC 2013). These accord CLAs with overall responsibility for the production of the chapter (IPCC 2013). There are usually two CLAs per chapter, and an attempt is made to divide this leadership role by assigning a developed country and a developing or EIT country author. The CLAs effectively manage the LA teams of 6 to 16 experts per chapter, depending on the WG, and maintain oversight over the chapter content and any cross-cutting issues between chapters. Lead authors are charged with writing given sections, as assigned during the first lead author meeting, and in preparing these they are encouraged to seek contributions from other experts in the field. They may also be requested to contribute to other chapters within or across WGs. Contributing authors are usually identified at the first and second lead author meetings and tend to be colleagues or members of the same academic networks as appointed authors. CAs do not attend author meetings; they are requested to submit technical information, such as text, graphs or data, which are then assimilated into the relevant section (IPCC 2013). When the nationality of contributing authors is incorporated in the breakdown of authorship, developed country dominance is further marked, with 90% of all CAs in the first four assessments belonging to Annex 1 countries (Ho-Lem et al. 2011: 1313). Review editors complete the chapter teams. Tasked with overseeing the chapter's review process and ensuring that all substantive review comments are given due consideration and assimilated into the chapter (IPCC 2013), these actors join the authors teams from the third lead author's meeting onwards.

The workloads of CLAs, LAs, CAs and REs require different levels of commitment and command different levels of influence over the chapter content. The IPCC indicates that the workload of the CLAs and LAs will be in the order of several months over the assessment cycle, with heavy periods towards the end of the drafting cycle (IPCC 2010i). The role of the CLA is most demanding. Responsible for overseeing the production of the chapter, these actors must ensure that the material submitted by the author team is arranged, harmonised and edited into a coherent chapter. This gives CLAs leadership and editorial power over the content, which increases with the time and institutional support – in the form of research assistance and technical and administrative resources – that are available to the individual to invest in the role (interview 1.08.2011). Although chapter visions are negotiated within and between the author team, and the ensuing text passes under many pairs of expert eyes at each stage of its development, it returns to the hands of the CLAs. Thus, just as the professional expertise of the WG co-chairs orientates the direction of the next assessment during the assembly of the outline, the CLAs' practice of the climate change problematic and epistemic connections shape their interpretation of the outline and the expert networks whose contribution and recognition are sought. CLAs are usually invited or volunteer for the drafting team of the SPM (see Section 7.1), thereby leading the process of identifying the chapter's key messages and conveying these in the most widely read constituent of the assessment.

For all participants, authorship requires a substantial time commitment over the course of the assessment. Authors are not remunerated by the IPCC, and participation is likely to reduce actor's research and publication output. As outlined in the letter requesting author nominations, developed country governments are expected to support the travel of those they nominate (IPCC 2010i), while the IPCC Trust Fund supports those from developing and EIT countries.[14] Through participation, IPCC authors expose themselves to criticism and personal attacks from actors seeking to undermine the organisation's credibility.[15] If the rewards are not financial, why do actors want to invest their time and professional expertise in the IPCC process? All participants interviewed were asked about the personal and professional benefits of contributing to the IPCC, and responses identified field-specific and IPCC-constituted interests. Field-specific interests are identified here, whereas those constituted in and through authors' investment in the IPCC's practice of writing are identified in Section 6.3, where authorship in practice is described.

The shared social interests of the authors of developed countries are largely constituted by the physical, natural and social scientific fields that qualify them as climate experts and which are the main audience for the reports. Being an IPCC, author enables actors to write about the field and their contribution within it, locating the major influences and advances in climate change knowledge, as well as remaining gaps. This assessment is subject to review by the wider field and will

[14] The IPCC Trust Fund covers IPCC activities, including participation of developing country participants at panel and bureau sessions, lead author meetings and other expert meetings. The fund also covers the cost of publication and translation of IPCC reports. This fund is maintained by voluntary contributions from member countries as well as contributions from WMO, UNEP and the UNFCCC.

[15] For example, those criticising the peer-review procedure followed in the SAR identified and held responsible, Ben Santor, the lead author of the section in question (Lahsen 1998; Skodvin 2000a; Edwards and Schneider 2001; Houghton 2008). Following the release of emails between IPCC authors at the Climate Research Unit at the University of East Anglia there were calls for the director, Phil Jones to resign (Monbiot 2010). Both cases are said to have had personal consequences for the individuals involved (Brown 2010). There is also evidence of American and Australian authors being subject to aggressive emails, abuse and even death threats (Bagley 2012; Butler 2010).

need to be redrafted accordingly, but in conducting and producing these overviews, authors secure their knowledge in the field. For climate scientists, atmospheric scientists and oceanographers of WGI, there is professional recognition in being appointed an IPCC author and considerable scientific authority attached to the CLA post. The awarding of the Nobel Peace Prize in 2007 further increased this cultural capital and extended it to the affiliated research institutes and universities, where the kudos of the Nobel Prize was incorporated into recruitment and funding strategies.

Authors of WGII noted that through participation they became aware of the limits and gaps in knowledge. Climate change impacts and adaptation are rapidly developing domains of study, and the IPCC has played a critical role in establishing the importance of this expertise to the UNFCCC negotiating process. As an IPCC author, actors have the opportunity to capitalise on the gaps they identify, either submitting publications before the end of the assessment cycle or creating grant proposals and new international research collaborations with the aim of generating knowledge for the next report (Hughes and Paterson 2017). These collaborations are valuable for both developed and developing country scientists, with an increased likelihood of success in grant capture because of the guaranteed platform for research outputs. In WGIII, interests depend on professional and disciplinary fields of practice. Interviews with economists suggested that IPCC authorship in the first two assessments was not recognised because scholarly contribution to the field was measured by publication in one or two major journals (interview 10.07.2010). More recent scholarship of AR5 WGIII authors reveals the dominance of economists and identified a group of authors who appear to have orientated their career and publication strategies around the IPCC assessment cycle (Corbera et al. 2016; Hughes and Paterson 2017).

Nationally, the levels of recognition given to IPCC authorship vary. Early studies of India's participation, for example, suggest that government funding agencies did not give as much value to lead authorship as North American and European governments (Biermann 1999: 8; Mahony 2014: 115–16). Whereas research in Brazil indicated that participation conferred prestige at the national level, which can result in lucrative consulting assignments with both national and international governmental and non-governmental entities (Lahsen 2004: 159). Authors from industry suggested that they became more knowledgeable of the climate field through participation and as a result more valuable to their clients and stakeholders (interview 19.09.2010). As well as field-specific and country-specific forms of interest, there appear to be perceived universal benefits to participation. These include networking with renowned experts in the field, transferring and expanding knowledge of climate change, making new friends and travelling to new countries (Tandon 2023).

6.3 The Order of Authorship in Practice

Following authors into lead author meetings and through the drafting cycle offers a clearer picture of how climate change is assessed in practice and the properties structuring these constructions. For the majority of actors, IPCC authorship begins with the receipt of a formal letter of appointment. Some will have been authors before, and a small percentage of those appointed will have contributed to the formation of the outline by sending in written comments and attending the scoping meeting, but for many this letter signals their initiation into the IPCC's practice of writing.[16] In preparation for the first lead author's meeting, the co-chairs request CLAs to lead the chapter team through the development of a more detailed outline. To assist authors in this process, TSUs provide detailed guidance notes. These notes provide background on the outline's production and summarise the discussions and concerns of governments that arose during the scoping and approval process, providing authors with a road map for the content of the chapter and indicating the political dynamics and points of contention surrounding the topics (interview 20.01.2011).

The first lead author's meeting is where the assessment takes form. This five-day meeting is when the chapter teams meet face to face for the first time, although in many instances, particularly in WGI and authors of multiple assessments, authors may be familiar with each other's work and even be well known to one another (interviews 27.09.2010; 30.10.2010). At this meeting, the chapter teams are charged with the task of finalising a detailed chapter outline, dividing the writing tasks between chapter members and devising an internal timeline for the preparation and compilation of the chapter (IPCC 2004). The schedule of the meeting is divided between plenary and chapter team sessions. The plenary sessions are the more formal of these, bringing authors in front of the IPCC chair, WG co-chairs and TSU staff to instil a sense of value in the IPCC and the assessment task (see Figure 6.2). These sessions are also the main avenue for inculcating authors with the IPCC's practice of writing: its processes and procedures, timelines and deadlines and relevant concept notes from expert meetings and workshops.[17] It is here that the WG co-chairs sensitivity and responsiveness to the political environment

[16] Percentages new to the process in AR5: WGI 65%; WGII 67%; WGIII 72%. AR6: WGI 61%; WGII 53%; WGIII 45% (IPCC 2018c, 2018d, 2018e).

[17] Expert meetings bring together a small group of experts identified by the WG bureau and TSU. Expert workshops are larger events and have a formal government nomination and WG bureau selection process, similar to that detailed in 6.1. There are usually a higher number of expert meetings and workshops in the early stages of the assessment cycle, which are designed to feed into the scoping process by tackling gaps and emergent areas of research since the previous assessments and by identifying the existent expertise and knowledge for the forthcoming report. These events, which often bring together communities of experts from across the three WGs, produce guidance papers and reports for authors. They also aim to initiate research collaborations and publications that will be available to assess in the forthcoming report.

Figure 6.2 The first lead author meeting of WGIII for the AR5 held in Changwon, South Korea. Photo by Emanuele Massetti: http://emanuele-massetti.blogspot.co.uk/2011/09/first-lead-author-meeting-of-ipcc-ar5.html.

and their role as conduit or interface between member governments and the scientific community becomes apparent, as during WGI's construction of the AR4.

WGI's report was co-chaired by US climate scientist, Susan Soloman. Susan was appointed co-chair at a time when the US administration was hostile to the science of climate change and the international negotiating process, as indicated by President Bush's rejection of the Kyoto Protocol and the government's attitude towards the conclusions reached by the IPCC's TAR and its chair, Robert Watson (McCright and Dunlap 2003, 2010). In 2001, the administration requested the National Academy of Science to undertake a review of the science of climate change, focusing particularly on 'where there are the greatest certainties and uncertainties' and 'whether there are any substantive differences between the IPCC reports and the IPCC summaries' (NRC 2001: appendix a). These events impacted the leadership of the AR4 from 2002 onwards, making them conscious that any conclusions reached in the report could be subject to congressional hearing. As a result, particular attention was given to the methods employed by authors to quantify certainty, to separate out fact from opinion and to ensure that there was a clear line of sight between the main report and the key findings in the SPM (interview

25.07.2010). The plenary of the author meetings is a time for the co-chairs and head of the TSU to share these concerns and ensure guidance notes for assessing uncertainties and non-peer-reviewed material are valued and adhered to in the writing of chapters.

Between plenaries, the WG break into chapter teams. The style and conduct of work are dependent on the WG and the academic and professional conventions of authors. The majority of IPCC authors come from government agencies and laboratories (~45% in the TAR and AR4) and universities (~40% in the TAR and AR4), with the remainder of authors from INGOs, NGOs or the private sector (Bhandari 2020: 197) Working group I is the most homogenous in disciplinary and diversity terms, while gender balance has increased from a low starting point; female authorship reached 27% of the author team for the AR6, compared to 40% in WGII and 31% in WGIII (IPCC 2019). There was an increase in the number of authors from developing countries and economies in transition in the AR6, which was up to 42% from 23% in the AR5 (IPCC 2018c), although the drop in US authors for the AR6 is a factor (Tandon 2023).[18] Diversity is an ongoing issue in the fields of science that contribute to WGI's assessment. Figures from the United States, which has been the largest contributor to IPCC authorship over the six assessment cycles (Tandon 2023), indicate that the lowest five occupational groupings for non-white-minority representation include 'atmospheric and space sciences, environmental and geosciences, and conservation and forestry' (Pearson and Schult 2014: 1040).

The cohesion in the authorship of WGI chapter teams means that the style of work, order of proceedings and social interactions – points of scientific contention and ways of contesting them – will be a familiar amalgamation of academic practice in the disciplinary fields that contribute to assessing the physical scientific basis of climate change. Less time spent establishing a means for conducting proceedings and negotiating a shared epistemology enables WGI authors to immediately burrow down into the content of the chapter, recent developments in climate science and the key messages that the chapter is likely to generate. However, it also means that working relations within WGI are most strongly governed by the dominant scientific order and its measures of scientific authority. This is an order with a propensity to marginalise and exclude the contribution of scientists that do not fit this embodiment.[19] In June 2018, a code of conduct was put in place for all IPCC meetings and events, as informed by UN and WMO instructions to prohibit discrimination, harassment and abuse of authority (IPCC 2018g). At the second lead author meeting, WGI CLAs, authors, bureau and TSU received training in

[18] The AR6 was the first time in an assessment that there was not a US chair of a WG. This was a period of hostility towards the UNFCCC and the IPCC under Trump's presidency.

[19] As depicted in the IPCC placeholder avatar, see Figure 7.2 in Standring 2022: 67.

inclusive practices, collaboration and participatory group dynamics as an attempt to broaden the space for participation (IPCC 2018f).

Although WGII is still largely composed of those practicing or managing research, its authors are more diverse in gender (40% in AR6), developing/EIT contribution (48% in AR6) and disciplinary make up, with chapter teams that span the natural and social sciences (IPCC 2018d, 2018f). This means that while authors continue to rely on shared academic practices for the conduct of their work, the substance and content of the chapters require negotiating disciplinary conventions, epistemologies and terminologies that on the surface are deceptively similar. This has become easier over subsequent assessments. During the compilation of the FAR, Dr Tegart, a WGII vice-chair, reported to the plenary on 'the complexity of the work' of WGII resulting from its multidisciplinary nature. He suggested that most of the experts involved 'had no previous interactions as they come from different disciplines' (IPCC 1990a: 20).

The IPCC has been an important facilitator of multidisciplinary collaboration, developing mechanisms to bridge epistemological divides and thereby deepen the level of engagement between disciplines through: (1) holding expert meetings and workshops that bring members of different communities together; (2) the treatment of cross-cutting issues and scenarios across working groups; and (3) developing concept papers providing authors with instructions and guidance notes to standardise assessment practice (interviews 25.07.2010, 5.10.2010). These guides on the treatment of uncertainties and the use of non-peer reviewed literature offer authors the means to assess the relevancy and validity of research that crosses disciplinary divides and for the treatment of these to be standardised across the WGs, although in practice, demands on authors' time can mean that these instructions are not fully integrated into the drafting process.[20] The AR6 cycle further deepened the integration between the WGs, with the Special Report on 1.5, the first produced by all three WGs. This established cross working group relations that were designed to continue throughout the assessment cycle (interview 26.02.2019; IPCC 2018c, 2018d, 2018e).

As with WGII, WGIII brings together authors from a variety of disciplinary fields, it is also home to more authors from IGOs, NGOs and the private sector (30% in the TAR and AR4), although authors from universities still dominate (40% in TAR and AR4) (Bhandari 2020: 197). Each of these professional fields has its own style of work and ways of understanding, producing and recognising knowledge of climate change; bringing these practitioners together in a single chapter necessarily gives rise to different dynamics and collective styles than in the other WGs. As a result, the order of relations may not be as structured by scientific

[20] For instance, authors are provided with guidance notes on the assessment of uncertainties and requested to produce a 'traceable account' of how they reached their expert judgments. Although a report by the Netherlands suggests that 'this part of the guidance has never been fully implemented in the assessment process' (PBL 2010: 31).

authority as in the other WGs, although the hierarchy of disciplinary knowledges, institutional affiliations and publication record remain enduring guides. Once the author meeting has broken into chapter teams, chapter members are able to size one another up. There may be some apprehension at first as chapter members find their place, but once proceedings are underway clashes of opinion surface and authors disagree over the chapter's direction, the material most relevant to the subheadings and the overall framing of the chapter.

The scientific habitus pervades the order of relations that emerge in the IPCC's practice for assessing climate change, structuring the space of relations and the extent an author and their knowledge is recognisable within and by a chapter team. Overlooked in this adherence to a social-scientific order is how an author's scientific capital to produce and contest climate knowledges is constrained and enabled by economic capital, level of English, race and gender (IPCC 1992b, 2018f; Gay-Antaki and Liverman 2018; Gay-Antaki 2021). Despite the fact that there are two CLAs, generally one from a developed country and one from a developing or EIT country, all interview data indicates that the developed country CLA leads the process. The authors that take the lead and whose voices are heard most in the decision-making and writing of the assessment appear on paper as the most accomplished in their contributions to knowledge and as such, it is only natural that they have the most to offer the production of the chapter. However, this natural scientific order overlooks and misrecognises the social, economic and material conditions that structure participation and capacity to imprint on the final product. A reoccurring theme during interviews, which is also evident in the Interacademy survey undertaken in 2010, is that not all appointed authors are adequately qualified and equally participate in the labour of the assessment.[21] For instance, one CLA notes:

> There are far too many politically correct appointments, so that developing country scientists are appointed who have insufficient scientific competence to do anything useful. This is reasonable if it is a learning experience, but in my chapter in the AR4 we had half of the LAs who were not competent. (IAC 2010b: 138)

While views like these are not explicit in recent surveys of IPCC authorship, concerns that balance overrides measures of scientific excellence in the appointment of authors persist, as indicated here

> … scientific excellence should still be by far the most important factor. I am extremely worried that the whole process could be seriously harmed if for the sake of balance in everything scientific excellence decreases. The imbalance need to be solved at universities, labs etc., and not at the IPCC level (IPCC 2018f: 28).

This perspective is mirrored by some developing country scientists' self-perceptions, as one LA comments:

[21] For comments that question the qualification of developing country experts and suggest 'tokenism' of participation see IAC 2010b: 138 and 16.

> The team members from the developing countries (including myself) were made to feel welcome and accepted as part of the team. In reality we were out of our intellectual depth as meaningful contributors to the process. (IAC 2010b: 330)

These attitudes generated by and generative of the social scientific order intersect with and overlook other determinants structuring the space for participation. Language and knowledge of the terms of the debate are cited as common barriers. When discussions become heated and the pace quickens, less-fluent English speakers have trouble following and are unlikely to interject in the proceedings (interview 7.07.2010c; Gay-Antaki and Liverman 2018: 2062; IPCC 2018f; Tandon 2021). Alongside language, gender and nationality intersect, as one African woman describes in a survey by Gay-Antaki and Liverman (2018):

> The only reason that I could have felt not required at all in the team could be that I am an African woman. I have very good command of English, I am as qualified as others, I am confident also—but I was never listened to. (Gay-Antaki and Liverman 2018: 2062)

The internalisation of this social-scientific order in authors self-perceptions leads authors to question their place and value, as an experienced author in global environmental assessments describes:

> You are never sure whether access to something is denied because of your colour, because of your gender, or because of your nationality. You're always questioning that. And I think the opposite is true as well – you aren't sure whether you were invited to something because they truly value your work or if they were trying to fulfil some diversity criteria. (Comment by Tuyeni Mwampamba, in Tandon 2021)

As authors begin to identify and assess the knowledge relevant to their assigned sub-heading, practical barriers to participation emerge, which again intersect and augment an author's capacity to invest in the IPCC's writing of climate change. Just as the acceptance rate for developing country scientists in international journals is much lower due to impeded access to current literature and perceived use of outdated methodologies (Gettelman 2003; Blicharska et al. 2017); limited access to international journals, slow and costly internet access, and even poor telephone connections impede developing country authors' capacity to access and assess relevant literature (Yamineva 2017: 28).[22] Schipper et al. (2021: 853) note, for example, that '... only a few African university libraries have reliable internet connectivity, with South African universities being among the most equipped in the continent'. Those leading the process have become aware of these barriers, and for the first time in the AR4 the WGI TSU reached an agreement with several publishing houses to provide authors with free access to journals. It was intended that this would be extended

[22] As one CLA notes, 'I had to send often articles to colleagues, notably African professors' (IAC 2010b: 618).

to all WGs for the AR5. However, WGIII was only able to offer a database and encourage sharing between authors. This meant that some developing country authors were again unable to efficiently search and access relevant literature and relied upon the support of other chapter team members (from correspondence with TSU). While this was meant to be resolved by a UNEP supported IPCC library facility for the AR6 (IPCC 2016a), issues with ease of access remained.

These asymmetries were amplified during the COVID pandemic and the dependence on virtual meetings it created. When the AR6 author meetings moved online, poor internet infrastructure meant that some developing country authors were 'cut off from the process altogether' (Julia Steinberger quoted in Ketcham 2022). Switching to virtual meeting arrangements made visible that the physical location or place of participation matters, and the time required to identify, access and review climate change knowledge is dependent on physical location, which the large hotels and conference venues of in-person author meetings obscure. Further confounding this divide is the fact that while many developed country authors, particularly CLAs, receive government or institutional support, which may include a research assistant for compiling lists of relevant literature and producing endnote databases, these funds are not available for developing country authors. The effect of this has been that while WGs have attracted developing country authors at the start of the process, authors have dropped out due to the time commitment as the assessment progresses (Yamineva 2010: 58).

IPCC authorship is demanding and time intensive. Although this labour is not financially rewarded, authors are able to distinguish themselves through their commitment to the process. One means of achieving this is through the figure. As well as reviewing, assessing and synthesising available knowledge through text, authors are keen to produce maps, tables, boxes, figures, and more recently, icons and infographics (Lorenzoni and Harold 2022). In a word restricted report, diagrams, graphs and tables allow authors to visually represent large volumes of data without taking up valuable space. These visualisations of climate knowledge are easily extracted for the SPM and from there, if they readily convey the extent and impacts of climate change, they may be taken into media and other forms of climate change reporting with the potential to influence international action. In the past, a few iconic figures have had a significant impact on popular discussion and negotiation, bringing attention (and criticism) to the IPCC and putting authors at the centre of controversy. Three important diagrams in this respect are the 'hockey stick' graph,[23] the 'burning

[23] The 'hockey stick' graph shows the average global temperature over the past 1,000 years and featured in Chapter 2 of WG I's contribution to the TAR and in the accompanying SPM. It has been updated in subsequent assessments. The graph caused dissent amongst the chapter team members and since its publication has become 'a symbol of the conflict between mainstream climate scientists and their critics' (Pearce 2010).

embers' diagram (Mahony 2014),[24] and the 'Bali Box' (Lahn and Sundqvist 2017).[25] Due to the advantages of diagrammatic representation and the attention it can bring to individual authors or chapter teams, the figure has become a significant object of interest, initiating contestation and competition as authors attempt to distinguish their contribution to the final product.

The IPCC also has its own rewards for those that demonstrate their proficiency through the conduct of the assessment and distinguish the value of their contribution and knowledge to the IPCC's practice of writing. Accruing IPCC specific capital in the eyes of those leading the assessment may result in being offered the opportunity to become more deeply involved in the current and next assessment through invitation to the drafting team of the SPM (Chapter 7) and the scoping of the next assessment (Chapter 6), and rising up the ranks of authorship and on to the bureau.

6.4 Reviewing

Turning from the authors to the review process makes apparent the collective nature of the IPCC's practice of writing. To ensure the emerging report is acceptable to both the fields of knowledge assessed and the governments accepting the reports, the IPCC has institutionalised an extensive expert and government review process. Nearly all scholarly text is subject to the approval of the field through some form of peer-review (Edwards and Schneider 2001: 229). The extended review procedures were not formally established for all IPCC reports until 1993 (IPCC 1993, appendix c annex 1). Today, reports are subject to three rounds of review, which typically leads 'to hundreds or even thousands of changes' in the text (Edwards and Schneider 2001: 235). The first review is conducted on the zero-order draft. This is when the assessment is largely at the stage of a list of topics and overview of relevant literatures. The document is reviewed internally and by a select number of external reviewers identified by the WG bureau, TSU staff and chapter authors (IPCC 2005, 2011b). At this stage, the reviewers are asked to consider the chapter's structure, gaps, balance, and cross-chapter issues, with the aim of providing authors with an early indication of whether the draft reflects the available literature and provides a balanced coverage of the chapter's scope (IPCC 2011b).

[24] The 'burning embers' diagram is figured in chapter 19 of WGII's contribution to the TAR (Smith et al. 2001) and the SPM. It identifies five reasons for concern, depicting the relationship between climate impacts and rising global temperatures. It travelled widely in media reporting. It has been updated for subsequent assessments (Zommers et al. 2020), although it has not always made it through the approval plenary.

[25] The 'Bali Box' in Chapter 13 of WGIII's contribution to the AR4 offered a quantification of the necessary emission reductions between Annex I and non-Annex I countries (Gupta et al. 2007). It became an important figure at COP 13 in Bali in December 2007, and later an object of struggle when the authors tried to update the figure by including reduction targets for developing countries (Lahn and Sundqvist 2017).

Comments from this review are discussed at the second lead author meeting and are incorporated into the construction of the first-order draft. Now the chapter begins to take shape and is made ready for its first expert review. Expert reviewers include those nominated as authors but not selected, experts put forward by relevant international and non-governmental organisations and individuals identified by the WG bureau and chapter review editors (IPCC 2021). For reviewers, the review process provides access to the IPCC's practice of writing and an avenue to influence how climate change is assessed. Special interest groups, such as environmental organisations and representatives of fossil-fuel industries, regard this as the main channel for transmitting their views into the assessment and for drawing author's attention to literatures beyond their purview (Yamineva 2010: 80).[26]

According to the IPCC's own calculations for the AR6, the total number of review comments across the three WGs exceeded 60,000 comments on the first-order draft (IPCC 2021). Many of these comments identify editorial errors or remark on the choice of topics covered by the chapter, which oftentimes authors have limited scope to address because of the government-approved outline.[27] During the roughly three-month review period, authors continue to develop the draft. This can make responding to the review comments a complex task, which requires, as one author noted, 'a tedious, confusing back-and-forth between the comment sheet, the formatted FOD and the active working draft' (Edwards 2022: 100). At this stage new literature is identified for review, and more substantial comments initiate chapter team discussion and debate as they assess the implications for the emerging second-order draft.

The review of the second-order draft returns it to the previous expert reviewers and, for the first time, exposes the emerging assessment to member governments that agreed its limits. As with expert review, the government review of the second-order draft is designed to gather comments on the accuracy, completeness and balance of the scientific and technical content of the draft reports (IPCC 2013), and for the AR6 this resulted in over 120,000 review comments across the three WGs (IPCC 2021). The review comments offer the co-chairs an opportunity to measure the reception of the report that governments sanctioned, with sufficient time before panel approval to address issues that comments identify as underdeveloped, missing or inappropriately formulated. As with author nomination, the internal particularities of the government review process depend on the focal point's location and how IPCC participation and climate change are institutionalised. For example, since the FAR, Australia has

[26] Agrawala (1998b: 626) suggests that some industry lobby groups have taken advantage of this in the past by submitting identical reviews on behalf of individual experts, certain non-governmental organisations and as part of the official US government review.

[27] Chapter teams can request to the bureau to amend one or two of the words in the government-approved outline.

held a national workshop to bring together experts from across different departments and outside of government to develop an agreed national view on the emerging assessment, which becomes a key document for the delegation during approval proceedings (Zillman 2008: 33). In earlier assessments, lack of trust in the scientists participating in this international collaborative exercise meant that the Brazil administration selected reviewers for the process (Lahsen 2004: 165).

All review comments are attributed; thus even if the reviewer is not known by name to the chapter team, the epistemic authority of the individual can be measured through national and institutional affiliation, which impacts how comments are received and dealt with by chapter team members. The overall response of the chapter teams to the review process depends on the actors and the attitude towards work cultivated, which also reflects the values instilled by those leading the process. Reports on errors made in regional chapters of WGII's assessment identified variation in the conscientious of chapter teams (IAC 2010a; PBL 2010). While in some instances it is necessary for review editors to reassure authors that review comments are unfounded (interview 10.11.2010), in other cases, lack of consideration for reviewers and their comments has left mistakes uncorrected.[28] This process of reviewing and revising in response to comments again highlights the asymmetries between developed and developing country participation in the IPCC's practice of writing.

The IPCC has undertaken its own study of developing country participation (IPCC 2009n, 2016b). This information gathering reveals a strong relationship between the nationality of authors and reviewers. In the data collected from the SAR to the AR5, developing country experts averaged 28% of the total authorship of the reports (CLAs, authors and review editors), and just 13% of reviewers (IPCC 2016b). WGII's contribution to the AR4 is one of the most widely reviewed reports, with 95 countries submitting comments. However, out of the 1,162 reviewers participating, 46% (529) came from five countries: the US (18%); UK (11%); Canada (7%); Australia (6%); and Finland (4%). In fact, across the three WGs of the AR4, four countries (US, UK, Canada and Australia) provided 43.7% of the authors and 41.1% of reviewers, see Table 6.1. For the AR5, developing country reviewers dropped from 15% to 11%, the lowest for which data is available (IPCC 2016b). This has a double impact. First, the content of the report is not subject to the questions and comments that help to ensure its relevance to developing country perspectives. Second, through conducting this review member governments gain deeper knowledge of the content and form a position on the text, ensuring informed and active involvement during the approval of a report's key findings (see Chapter 7).

[28] In the case of the erroneous statement in WGII's assessment on the melting of Himalayan glaciers, two reviewers commented on this during the government review, one of which provided references for articles that drew different conclusions to the chapter (IAC 2010a: 22).

Table 6.1 *Number of authors and reviewers by the top five countries for WGII's contribution to the AR4*

Country	Number of authors*	Percentage of total (%)	Overall ranking	Number of reviewers	Percentage of total (%)	Overall ranking
US	73	15.8	**1**	209	18	**1**
UK	60.5	13.1	**2**	128	11	**2**
Canada	38	8.2	**3**	76	6.5	**3**
Australia	31	6.7	**4**	65	5.6	**4**
Finland	3	0.6	**17**	51	4.4	**5**

* Author count includes coordinating lead authors, lead authors, contributing authors and review editors. All figures are based on author's own calculations from list of authors in Parry et al., 2007.

6.5 Re-ordering Author Relations

The IPCC has sustained considerable criticism over the diversity of participation. As this chapter demonstrates, it is a simple exercise to illustrate the dominance of men over women and a small number of countries over the IPCC's practice and authorship of writing climate change. In response to this criticism, the co-chairs, bureau and wider panel have undertaken their own information gathering and taken steps to increase gender balance and diversify and deepen developing country participation in IPCC activities (IPCC 1992b, 2009n, 2016b, 2018f, 2018h, 2019). Participation, however, does not guarantee meaningful impact on the organisation and its writing of climate change when, as documented, there are resource and institutional constraints on the capacity of authors to invest, and when the social order overlooks the knowledge and authority of developing country experts and the asymmetrical resource distributions that structure scientific careers and participation in the process.

The criteria for selecting authors, the standardisation of the assessment practice and, more recently, diversity training disrupt dependence on the scientific habitus to identify climate expertise and to order relations in the conduct of the assessment. Although measures of scientific excellence remain pervasive, the AR6 was the most diverse assessment cycle in terms of gender and developing country participation. However, increasing the number of women in the assessment may prove easier than continuing to increase and maintain developing country participation and certainly easier than shaping the social order that would enable meaningful participation by *all* authors regardless of nationality, gender, race, publication record and institutional affiliation. As this chapter describes, the asymmetries in the global knowledge economy remain, and even those aspects of this that the IPCC would seem to have power to effect, such as access to literature for all authors, are proving difficult for the organisation to adequately address. Material factors intersect with language, gender and race and are generative of the order of relations through which climate change is written in and through the IPCC's assessment practice. As captured in the IPCC's own analysis:

> For several questions, the strongest differences in responses between men and women occurred in the Latin American/Caribbean region. Women were three to more than eight times more likely than men of that region to give a negative response regarding gender balance and bias in the IPCC. Women were also up to nine times more likely to give a negative response than their male colleagues from that region on their experiences with the IPCC. (IPCC 2019: 26)

Experiences of participation currently diverge, and it cannot be assumed that increasing the number of white women in the assessment and in leadership roles – as important as this is – will necessarily result in broadened inclusivity for all authors (Dhillon 2020, 2022). The organisational attentiveness to diversity and order, which include and expand context specific evaluation of nominees, diversity and unconscious bias training, assessment skills workshops and the capacity to report and reshape relations during the assessment will continue to be critical in the IPCC's journey in creating a more representative and fairer assessment practice for naming climate change.

6.6 Summing Up

The order of authorship in the IPCC's practice of writing is not what it was when the organisation was established in 1988. During early assessments, practices for nominating and appointing authors relied on the scientific networks of those leading the process to identify and extend the author team. Mistakes in reporting, scrutiny and struggle over final wording by member governments and criticism in scholarship and by those contesting climate reality have resulted in increasing codification of authorship. This includes selection criteria to ensure geographical representation, a range of views, retaining experience alongside introducing new experts and gender balance in the selection of author teams. While scientific measures of authority continue to distinguish nominees and order relations within the chapter teams, the codification of the IPCC's assessment practice exerts its own force. Those leading the process have the means to instil in authors a code of conduct that upholds the values of the organisation and aims to retain its privileged position within the climate field through the assessment practice. The TSUs are critical to achieving this, preparing documentation that provides background on the outline's formation, producing concept papers to standardise assessment procedures, inculcating author dedication to the onerous drafting and reviewing cycle and ensuring that harassment and unfair treatment can be reported and acted upon.

Despite these tangible changes, the dominance of developed countries, or the global north, in the authorship, knowledge assessed, and review comments submitted remains. To understand this persistence, the chapter has explored broader practices and patterns of global knowledge production that structure the space for developing country experts to become internationally recognised climate scientists, as measurable by the TSU's metrics and acknowledged by and within author teams. This reveals the coupling between the global knowledge order and the global distribution of economic resources, which is masked and easily overlooked in and by the scientific culture that orders and organises the relations and practices of assessing climate knowledge. As Collins (2015: 2) describes in defining intersectionality, 'the devil is in the details', which is why it is important to unravel and describe each activity that constitutes authorship within the IPCC's assessment practice. Taking author nomination, the order of authorship, or reviewing as single separate stages identifies distinct ways in which the space for developing country participation in and contribution to global climate knowledge is pre-structured. Combine each of these activities and add the outline of the report that informed it, and the relationship between the maldistribution of resources and dominance over knowledge production is revealed. This is a vastly unequal global climate knowledge order.

Analysing each activity as an element within the practice of writing makes another, related pattern, discernible. Each activity in an assessment's production enables governments to involve and invest in this global attempt to write climate change, ensuring the product is relevant to the national interests and needs for and from climate knowledge. Nominating authors and conducting a government review of the emerging text are not simply avenues to shape the content; they are avenues to know the content – to learn the latest knowledge on climate change, assess and disseminate it across government, develop an informed position and ensure the final product is relevant to national climate policy needs. It is those countries that have the resources to invest, fulfil the government activities and gain knowledge through their undertaking that the IPCC's practice of writing best serves. It is with this in mind that I move from the scientific assessment to government approval of the report's key findings in Chapter 7. It is here that these patterns of dominance can become exposed and struggled over as governments attempt to ensure that climate change, as written in and by the SPM, is relevant to and supports their order-making in its name.

7
The Politics of Approval

This is the final stage of producing an IPCC assessment report on climate change – the last chance to shape its writing, and in its most significant form. The summary for policymakers (SPM) presents in sentences, bullets, boxes and figures the findings that are most relevant for social and political decision-making from the underlying assessment report. They are the most circulated IPCC products, bringing the key messages to the surface, ready for dissemination and effect. Here sentences and figures travel into minister's speeches, media coverage, government, UN and NGO reports, where they shape negotiation of and policymaking on climate change domestically and internationally (Hermansen et al. 2021). It is therefore in the approval session that the extent of political struggle over climate change in the IPCC is most visible.

The attempt to shape the writing of climate change began with the election of the bureau and importantly, with the approval of the outline, which as documented in Chapter 5, brings to light government attempts to delimit how climate change is written in the next assessment. The nomination of authors and the government review of both the full report (Chapter 6) and the emerging SPM are also important avenues for influencing the construction of climate change. However, once the final draft SPM is delivered to member governments, often a day or even just an hour before the WG approval session opens, the last opportunity to contain, control or elevate the meaning and implications of climate change, as written by the IPCC, begins. These are the stakes in the approval of this document.

Despite this document's importance, in relation to other aspects of the assessment process, the IPCC's 'line-by-line approval' is relatively under studied. As scholars have gained access to and observed these intergovernmental sessions, they have identified the amalgamation of scientific and political activities (Shaw 2000, 2005; Petersen 2006) and conceptualised these as forms of consensus building (De Pryck 2021, 2022), uncertainty management (Fogel 2005), and simultaneously, both an attempt to create a single climate story for political action (Livingston, Lövbrand

and Olsson 2018) and as a method for pluralising it (Kouw and Petersen 2018). However, to grasp fully what is at stake in this practice of approving the key climate messages, it is critical to situate the IPCC within the international field of climate action and to bring its relation to the United Nations Framework Convention on Climate Change (UNFCCC) into focus (Hughes 2015). This makes it possible to identify the stakes in the practice of approval, the forces that the politics of climate action generates and their imprint on the final SPM product.

In describing the order of relations in the panel (Chapters 4 and 5) and the scientific assessment (Chapter 6), I have chartered the dominance of the global north in the IPCC's practice of writing. In this chapter, in and through recounting the politics of approval, I illuminate how the imprint of these asymmetries on the naming of climate change are challenged. At the outset, describing the drafting and reviewing of the SPM and taking a closer look at participation in approval sessions, the order looks much as it did from the outset, although the emergent scientific strength of some countries is apparent in the authorship of the AR6. When the final draft passes from the hands of the co-chairs to the delegates, the approval of the key messages becomes the site for developing countries to contest the framing of climate change that the global north's scientific dominance writes. I use the struggle over country categorisation, which reached its height during the approval of WGIII's contribution to the AR5, to describe the strategies available to the co-chairs, delegates and authors to influence proceedings. This amounts to a politics of approval that can overflow the meeting, as authors, co-chairs and delegates carry their wins and carefully crafted sentence or frustrations and lost text for reuse in their broader participation in the science and politics of writing climate change.

7.1 The Stakes in the Practice of Approval

The idea of producing a 'policy document' to summarise the scientific results of the full report is said to have been a WGI proposal accepted by the IPCC bureau at its first session in February 1989 (IPCC 1989 in Agrawala 1998b: 633). There had been an intention to generate some form of policy document from the outset. Initially, it was envisioned that the bureau would integrate the three WG reports of the first assessment report (FAR) into 'a single assessment statement which will include an executive summary, designed to facilitate the requirements of political analysts' (IPCC 1988: 6). The first version of this was produced by WGI for the FAR and followed a comparable route to the pathway now embedded and described in Section 7.2.[1] The text was subject to approval at a three-day meeting attended by the authors, other invited experts, delegates from 35 countries, environmental

[1] See Houghton 2002: 3.

NGOs and representatives from oil and coal industries (Leggett 1999). The meeting has been described as collegial and the criteria used for the document's approval scientific (Lunde 1991: 82; Houghton 2002). WGII and WGIII were also requested to produce a policy document. At this stage in the IPCC's development, the assessment practice had not been standardised across the three WGs (see Sections 4.6 and 6.3). Whereas WGI could rely on scientific conventions to structure the order and conduct of its assessment activities, the varied professional and disciplinary composition of WGII and III meant those involved did not have a shared habitus or related conventions to order the conduct of proceedings. This, along with the political nature of the content, meant that producing an assessment and summary of the impacts (WGII) and response strategies (WGIII) was a more troubled process (Hecht and Tirpak 1995: 385–86; Skodvin 2000a: 119–23; Bolin 2007: 63–66).

With the completion of the three WG reports of the FAR, the IPCC chair, Bert Bolin, prepared a synthesis report to highlight the key findings from across the assessments for approval by member governments (IPCC 1990c; Bolin 2007: 67). The chair's document did not make it through the plenary, however, and it seems that it was only possible to reach closure 'by cobbling together' lengthy extracts from the WGI SPM 'and a few of the less contentious conclusions from the politically sensitive WGII and WGIII reports' (Zillman 2007: 879). The proceedings of this session were characterised by political dynamics that have become a recognised feature of climate negotiations.

> Having started in a very civilised fashion with songs about the future from children's choirs and an address from the prime minister of Sweden, the meeting finally came very close to breakdown. It finished at four o'clock in the morning, one day late, with most of the delegates having abandoned their chairs in the conference hall to gather on the front podium and shout at each other. (Brenton 1994: 183)

It is clear from this account that emerging national positions on the issue of climate change shaped government attitudes towards the drafted text (Hecht and Tirpak 1995: 386–87). For instance, the American delegation wanted the uncertainty of the science emphasised (Lunde 1991: 82; Leggett 1999), the former USSR wanted caveats added and possible benefits to agriculture highlighted (Lunde 1991: 96; Hecht and Tirpak 1995; Leggett 1999: 15–16), the Brazilian delegation arrived with a new study that contested the report's depiction of the contribution of tropical deforestation (Lunde 1991: 97) and other developing country delegations refused to join a consensus document (Brenton 1994: 182–83; Zillman 2007: 879).

Observer status to IPCC meetings also gave access to lobbyists from fossil fuel industries and environmental NGOs. At this stage, the role of non-governmental actors was 'loosely defined', and they were permitted to intervene and make suggestions for the wording of the text (Leggett 1999: 3). As a result, these actors also became part of the struggle as they attempted to insert their interests into the policy document. This participation was restrained during the approval of the SAR, when the co-chair of WGI requested observer organisations to leave the floor to governments (Lunde 1991: 77–78; Leggett 1999: 229–30). The tensions that arose during the approval and finalisation of the FAR have become a permanent feature of the line-by-line approval of IPCC documents, and to the observer, contiguous to UNFCCC negotiations (Hughes and Vadrot 2023). To understand why the approval has become a recognisable site of struggle in the climate field, it is necessary to bring the IPCC's relation to the UNFCCC into focus and to examine the role that the IPCC's assessment practice and its knowledge products have on climate negotiations.

In a study of the Special Report on Land Use, Land-Use Change and Forestry (LULUCF) (IPCC 2000), Fogel begins to unpack how the IPCC can and has been used as a site for 'legitimating and refining' political outcomes negotiated within the UNFCCC (Fogel 2005: 206). The LULUCF special report was requested by SBSTA in June 1998 to assess the state of scientific and technical understanding on carbon sequestration in response to controversial policies agreed in the Kyoto Protocol (IPCC 2000). Through observational research of both the UNFCCC negotiations and the IPCC approval (Fogel 2005: 193), Fogel's study reveals how the special report provided the scientific basis and method for operationalising policies and decisions that were crafted in the Kyoto Protocol 'for political and economic reasons' (Fogel 2005: 206). The effect of using the IPCC assessment practice for this purpose is to bring or extend UNFCCC negotiations into IPCC approval sessions, and Fogel's article documents the extent of political manoeuvring and bargaining over the LULUCF report's key findings.

While Fogel's article identifies how the IPCC's assessment practice can serve to legitimate negotiated decisions, a study by Lahn and Sundqvist reveals the potential for IPCC knowledge products to inform and shape the negotiation of decisions (Lahn and Sundqvist 2017). The authors of the study follow a figure, the so-called Bali Box, from WGIII's contribution to the AR4 into negotiations towards the development of a post-Kyoto framework at COP 13 in Bali. The figure provided a proposal to a key sticking point – a quantified distribution for equitably sharing the burden of emission reductions between developed (Annex 1) and developing countries (non-Annex 1) (Gupta, Tirpak, Burger 2007: 776). Although the approach offered in the Bali Box was ultimately rejected, its travel from the assessment into the negotiations and role in shaping discussions demonstrate how IPCC assessment products feed into and tangibly shape collective decision-making in the UNFCCC.

Since these initial studies, the evidence for the IPCC's role in producing, modifying and legitimating objects of negotiation has grown,[2] with the Special Report on 1.5 offering the clearest example of this. Through the UNFCCC's invitation and the production and dissemination of the report, the 1.5 temperature goal, which was not a significant

[2] On the notion of the 'carbon budget' from the AR5, see Lahn 2021; Coppenolle, Blondeel and de Graaf 2022.

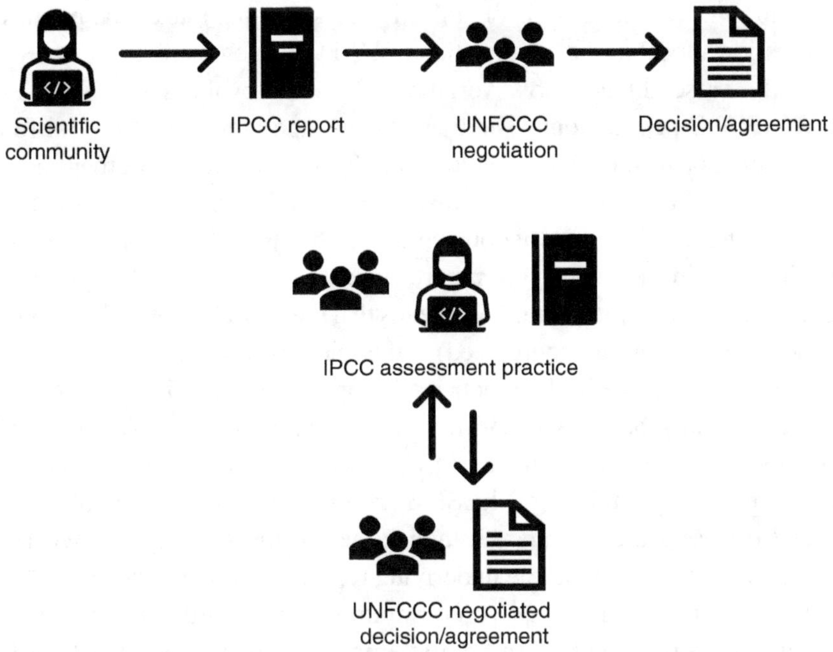

Figure 7.1 Top: linear model of how science influences political decision-making; bottom: observed circulation of actors and objects in IPCC practice of producing assessments and UNFCCC sites and processes of negotiation.

object of scientific interest before the Paris Agreement (Livingston and Rummukainen 2020), became a legitimated target of study in the scientific community and negotiation in the collective response (Tschakert 2015; Livingston and Rummukainen 2020; Beek et al. 2022). These studies make apparent that the origin of constituent objects of negotiation and agreement formation – such as the 1.5 temperature goal – do not follow a linear pathway from the scientific community to the IPCC to UNFCCC negotiations (Figure 7.1). Instead, objects pass between the IPCC and UNFCCC with the actors (delegates, bureau members and authors) and the products (reports) and outcomes (decisions) they produce (Figure 7.1). These objects may appear to originate from authors and the broader scientific community, as with the Bali Box, or from parties and negotiations, as in the LULUCF and 1.5 special reports, but are most likely some combination of the two.[3] In this respect, central objects of climate science and politics are continually taking shape as they are assessed and approved in the IPCC and negotiated in the UNFCCC and/or in reverse, in a process that has been described as 'mutual validation between these worlds' (Van der Sluijis et al. 1998: 315). It is this – the stakes of making

[3] Whether the object appears to originate from the scientific literature or a negotiated decision, they are already hybrid/co-produced objects because of how the negotiation and institutionalisation of climate politics influences climate knowledge production and how climate knowledge informs political decision-making nationally and internationally (Jasanoff 2004a; Miller 2004; Hughes and Paterson 2017).

(or un-making) the scientific basis of collective action – that makes the politics of the approval session appear as the continuation of UNFCCC negotiations.

Despite their resemblance, however, the IPCC's approval practice is not a carbon copy of UNFCCC negotiations. There are three important factors that differentiate the practice of approval: (1) the SPM text and the assessment report that underlies it; (2) the authors; (3) the WG co-chairs that oversee the report's production and chair its approval. The SPM is not a negotiated text in the way that a UNFCCC decision is; it has its basis in the underlying report, which has its basis in the published and peer-reviewed scientific literature. This means that there is a line of sight to its creation and an anchor (real, crafted or perceived) in the scientific community. As described in relation to the authors role in the practice of approval (Section 7.3.3), this constrains what can and cannot be revised and how it is re-phrased and re-written, although this does not always prevent sentences from becoming vaguer, more ambiguous or deleted altogether. The authors of the SPM are the designated judges of this as they present and represent the science – establish the anchor – and accept whether proposed revisions are in line with the underlying report (interview 5.08.2010).[4]

The chairs sit between these two communities – the authors and the member governments – deeply invested in both attempting to guard the report's key messages and reaching a government-approved SPM document. This is no small feat. The pressure on the approval process has grown as the global struggle over climate change has intensified and as member governments increasingly recognise and strategically use the IPCC as a site of negotiation, bringing UNFCCC negotiations into the IPCC's practice of approval. The significance of the IPCC as a site in and of climate agreement-making is likely to continue to increase as a result of the Paris Agreement, which effectively tasked the IPCC with providing the 'best available science' to evaluate the agreement's implementation (UNFCCC 2015). Before exploring how the dynamics between the authors, co-chairs and member governments shape the practice of approval, I describe the activities and politics of preparing an SPM.

7.2 The Order of Drafting and Reviewing

The codification of the line-by-line approval of the SPM emerged through subsequent iterations of IPCC rules and procedures. In the principles governing IPCC work that were agreed after the FAR, consensus was specified for the conduct of the meeting: 'the IPCC Plenary and Working Groups shall use all best endeavours to reach consensus', and in cases where this is not possible, 'differing views shall be explained, and, upon request, recorded' (IPCC 1991: 8). Further codification

[4] The emphasis here is on the role of the authors in establishing the anchor; on anchoring devices see van der Sluijs et al. 1998.

of these rules followed through periodic review, with a subsequent iteration stating that 'Reports approved by the Working Groups and accepted by the Panel will principally be the three Executive Summaries and the three Summaries for Policymakers.... The Summaries should be subject to review by both experts and governments and to final *line-by-line approval* at a Plenary meeting of the appropriate Working Group' (IPCC 1993, appendix G, italics added). This codified the practice of approval, introducing a review of the emerging draft that would enable governments to comment on the report's key findings and to prepare for the session.

The incremental development and learning required in realising this approval in practice, however, was evident when the WGI co-chair, Sir John Houghton, arrived at the approval session for the FAR with a 40-page document. Despite protests from colleagues that an SPM of that size would never make it through, the proceedings got underway and became mired in controversy, as delegates took to contesting every line and in some instances every word (interview 9.11.2010). Eventually, these 40-pages became the first ever technical summary, and the executive summary to the report was converted into the SPM (Leggett 1999: 227; Skodvin 2000a: 215). The history of the emergence of the SPM, the pathway for its construction and the codification of rules for conducting the plenary approval session indicate the scale of the task that the IPCC set itself in aiming to produce a policy document that required both a practice for achieving it and a shared value in realising a collective knowledge base for negotiating climate action.

Although there is a clear pathway for producing an SPM, it is not fixed. As with the assessment reports, there are institutionalised procedures that have to be followed: a page limit, a government review and an order and timeline for re-drafting. The particulars of the process – the selection of the drafting team and the message to be conveyed – are dependent on the assessment round as directed by the co-chairs and as situated in the political context at the time. It is the WG co-chairs that have overall responsibility for preparing the SPM (IPCC 2013), and formally, the process for selecting the core writing team is a decision of the WG bureau (IPCC 2005: 2). In practice, the drafting team is assembled through discussion and feedback from the TSU, wider bureau and chapter team authors (interview 20.01.2011).

The convention is to have two representatives from each chapter, either both chapter CLAs or a CLA and a lead author. For those assembling this drafting team, either at the level of co-chair or within the chapter teams, there are practical concerns and anticipation for the final destination and potential contestations to be considered. During the intergovernmental session, authors present and explain the scientific findings that underpin the report's key messages, and therefore the WG co-chairs need a drafting team that can facilitate the SPM's travel through member government's line-by-line scrutiny. From the perspective of co-chairs and the TSU, authors need to have demonstrated the capacity to deliver and to have displayed attributes, such as the 'ability to summarise and think clearly' as required in the final approval (interview 14.07.2010). Confident language skills become an important consideration, as a 'shift' in the way the science is spoken is necessary for presenting it to member governments (interview 7.07.2010c).

For authors, drafting the SPM will be an additional workload to the already pressing demands of IPCC authorship, and they need to volunteer themselves for this intensive role. There is, however, prestige in being a member of the core writing team. The authors are aware that in most cases the only people reading the entire chapter are reviewers, peers and students in the field (interview 7.07.2010c). Being part of the SPM writing team, on the other hand, provides authors with the opportunity to work on a widely read document that is influential in social and political constructions of climate change, for which they are prominently credited at the front of the document. These authors will work closely with those managing the assessment process, will be invited to additional drafting meetings, will participate in the approval session (Schneider 2009; Stavins 2014; Broome 2020) and may become key actors in the report's dissemination at the national and international level (interview 26.06.2023). This includes being invited to present the key findings at UNFCCC mandated events, expert dialogues and official and unofficial side events.

The writing of the SPM remains in the hands of a relatively small group of countries. Across all three WGs, 60% of the drafting authors for the AR5 and AR6 came from 14 countries, as shown in Figure 7.2. Of this, three countries made up over one third or 34% of the authors: the US (14%), Germany (10%) and the UK (10%). All of the top 14 countries, except the Netherlands, had a co-chair in the assessment or a member on the bureau, and government support for authorship is also a factor.[5]

The emerging SPM undergoes an expert and government review, and the comments are discussed and responded to by the drafting team during the fourth lead author's meeting (IPCC 2006c). Once redrafted, the SPM is sent out for a final government review before being finalised for approval. The government review is identified as an opportunity to improve the document, to make sure that the content covers the most policy relevant issues, without being policy prescriptive, and that the language is appropriate for a policymaker audience and consistent throughout the report. Many of the comments submitted are about the general presentation and structure of the summary, the use of technical language and inconsistencies in terms and parameters. The reference point for these comments is frequently the SPM of the previous assessment, with the current product checked against the clarity and conclusions of the previous text. Carried through these comments is the distinct view of government actors and their

[5] The Netherlands co-chaired WGIII for the TAR and AR4.

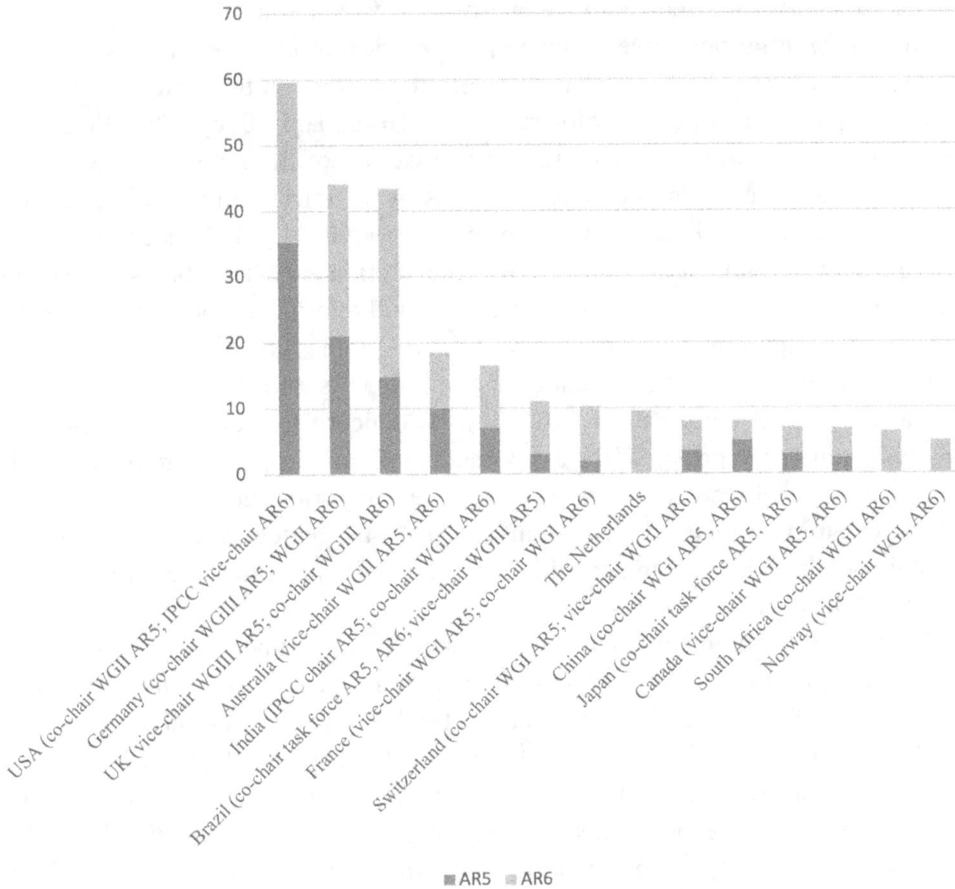

Figure 7.2 Top 14 countries by number of SPM drafting authors from WGI, WGII and WGIII in the AR5 and AR6.

interests in climate politics (see Table 7.1). Many of the government reviewers providing input have first-hand experience of UNFCCC negotiations, extending into participation in negotiation of the decisions and agreements reached over time as reviewed in the assessment, and are therefore quick to identify author's misunderstandings or mis-formulated descriptions of the process (IPCC 2014a). These comments often reveal the national position on the text and signal to the co-chairs and the authors the topics, phrases and words that will excite the most debate during the approval.

Assessment of countries into developed and developing categories and discussions of historical versus present and future emissions are central issues of struggle because of their bearing on responsibility for emission reductions in the UNFCCC. In the Kyoto Protocol, guided by the principle of common but differentiated responsibilities or CBDR, Annex I or developed countries took the lead on quantified emission reductions. However, the principle of CBDR and the differentiation between developed and developing country responsibilities for emissions reductions that it underpins were open to reinterpretation in the development of a post-Kyoto framework (Rajamani 2016). The IPCC was and is situated centrally in this struggle as assessor of the knowledge base and methodologies for evaluating and categorising countries and GHG emission reductions. This was evident in Lahn and Sundqvist's (2017) account of the struggle over the Bali Box from the AR4 (see Section 7.1). The stakes were even higher in the AR5, which was being prepared and approved alongside the negotiations, in which country grouping was 'perhaps the single most contentious issue' because of its linkage to national commitments under the new agreement (Dubash, Fleurbaey and Kartha 2014 36). Its concern to member governments was evident in the approval of the assessment outline, as described in Section 5.4. While many developed countries wanted these categories subject to assessment, the larger emitters resisted any attempt to open this categorisation to analysis through the identification and specification of chapter headings and bullets in the report outline (Section 5.4).

This struggle continued across government review comments of the final draft SPM, as demonstrated in Table 7.1. Although, in most cases the country name has been removed from the collated comments (IPCC 2014a), it is possible to identify the distinct views and perspectives of developed versus developing countries in the comments (see Table 7.1). In general, developed countries stress the lack of clarity in authors' categorisations and request present and future emissions to be emphasised over past, while developing countries stress the need to distinguish historical emissions and highlight to authors the differentiated commitments of developing countries.

In most instances, the governments participating in the review and submitting the most detailed comments are the same countries that make up the majority of the authorship of the reports. In the AR4, for example, which had the highest number of developing country reviewers at 15% of the total number of countries submitting comments (IPCC 2016b), the ten countries providing the most commentary on the first draft of WGI's SPM were the US, Canada, UK, Germany, Australia, Norway, Austria, France, Japan and China.[6] Nine of these countries (excluding Austria) accounted for 78% of the authorship of the SPM, and of the forty countries contributing to the full WGI report, these ten countries constituted 86% of the CLAs, 68% of the lead authors and 62% of the Review Editors, with the US and the UK combined accounting for over one third of the authorship of the WGI

[6] These calculations are author's own, made from the record of government's comments on the first review of the SPM (IPCC 2006d).

Table 7.1 *Government comments on the final draft SPM of WGIII's contribution to the AR5.*

Annex I country positions*	Non-Annex I country positions
The SPM largely neglects changes in the balance of **developed/ developing** world emissions shares – for instance, entirely omitting the (recurrent) Chapter 7 headline that Asian coal consumption has substantially driven escalating global emissions, since at least 2001 (see, e.g., p. 11, l. 7–16 and p. 14, l. 21–28) (IPCC 2014a: 125).	We request to delete this phrase: "and because the **Kyoto Protocol** does not directly regulate the **emissions** of **non-Annex I countries**, which have grown rapidly over the past decade" (IPCC 2014a: 117).
CHAPTER 3, P. 12, LINES 16–22: The statement that, "**developed countries** bear much of the causal **responsibility** for climate change because of their **historical emissions**" is biased and factually incorrect, given that **developing country** cumulative **emissions** during 1850–2010 make up 48% of global GHG **emissions** (with **developed country emissions** accounting for the remaining 52%) (den Elzen et al. 2007 Climatic Change). Another way of making a similar point would be to say that countries' **historic emissions** help determine their causal **responsibility** for climate change, without making the **developed vs. developing country** distinction (IPCC 2014a: 23).	The Government of China … The SPM should present the overall picture of **historical** status in flows and stocks of greenhouse gases (GHGs) **emissions** and their drivers in a balanced and comprehensive manner. However, the SPM fails to provide the conclusions and information on stocks of global GHGs emissions and per capita **emissions**, but just highlights global GHGs flows of recent years in a selective manner. For example, Section 2 of the SPM merely emphasizes near-term **emission** status from 2000 to 2010 which is highly sensitive to its starting and ending years, but neglects more important long-term emission trends. In addition, the SPM only analyzes total accumulative amount of CO_2 without considering the population factor. It is suggested to reflect global GHGs **emissions** of different time spans in a more comprehensive and balanced manner in the SPM that includes flows and stocks, total and per capita **emissions**, **emission** increases in absolute and relative terms, and global aggregation and regional distribution, in particular the information on **historical** per capita accumulative **emissions** in the RC5 region set (IPCC 2014a: 1).
Since the statement covers 1750 through 2010, the authors should note the much altered (and still changing) composition of the major-emitting country group – the **developing** world having overtaken **historically** higher **developed country emissions**, and the "small number of countries account[ing] for a large share of global CO_2 **emissions**" (IPCC 2014a: 26). Regarding the statement "… how to account for such factors as **historical responsibilities** for **emissions** …", suggest also adding "and anticipated future contributions to **emissions**" in order to account for both forward and backward-looking perspectives in this list of examples of factors associated with ethical questions (IPCC 2014: 135).	General comments introduction: It is important to emphasize the issue that climate change is contingent upon the fulfillment of committments of **developed country** Parties based on the historical responsibility and the principle of **Common but Differentiated Responsibilities**, and ensuring financial, technological and capacity building for the necessary climate adaptation (IPCC 2014a: 123). The assumption that "all countries of the world begin mitigation immediately" is policy-prescriptive in that it suggest that **developed countries** and **developing countries** mitigate in an undifferentiated manner. If most studies indeed make this political assumption, then qualifying language is needed to make clear that matter of **CBDR&RC** in the future climate regime is still open in negotiations, and that the literature reflects differing views on this (IPCC 2014a: 49).

Table 7.1 *(cont.)*

The use of such broad regional groupings (e.g., "**developing**", "**developed**", **annex 1**, **non-annex 1**) does not provide useful information to policy makers. It would be informative to provide information on major economies to avoid making sweeping generalisations and loss of detail (IPCC 2014a: 17).	The overall objective of reducing greenhouse gases and timeframe thereof must be based on **historical responsibility** and the full implementation of the **Convention** (IPCC 2014a: 110).
The statement regarding per capita **emissions** being "markedly higher" in the **Annex I** group is not consistent with Figure 1.8(c), which shows that many **non- Annex I** nations (including South Africa, Brazil, China, Saudi Arabia, Indonesia, etc.) have per capita emissions on par with **Annex I** countries. As a result, the second part of this sentence needs to be deleted (IPCC 2014a: 21).	This statement lack comprehensive **historical** assessment and provides insight to an extremely short period of time. A comparison with corresponding **historical** trend such as **emission** patterns of past major economies is required for policy makers to grasp the broad analysis (IPCC 2014a: 12).
I think that, in a globalized world, "countries" may not be the best, or at least the only, basis of concluding who is more responsible for emissions etc. (For example, rich and poor people; regions within and between countries; or multinational companies could be used as groups.) I suggest to at least mention that the current use of countries as **categories** are but one possible approaches, and other **categorizations** might be useful to identify mitigation policies and measures. (IPCC 2014: 26)	**Protocol** does not directly regulate the emissions of **non-Annex I** countries, which have grown rapidly ~ **Protocol** does not directly regulate the emissions of THE GROUP OF the **non-Annex I** countries, which have grown rapidly (the very rapid growth is valid for a part of the **non-Annex I countries**, therefore the rapid growth is valid only for the group, but not for all **non-Annex I countries**) (IPCC 2014a: 16).
It would be helpful to define which countries fall in the income-level categories used in the SPM (e.g high income, upper middle income, lower middle income, and low income). It would be useful to have this within the SPM itself or through a link to the definition in the appendix/glossary. The definition should include the ability to view the specific countries listed in each **category** in addition to the income brackets (IPCC 2014a: 124).	This paragraph must include the concept of the **historical responsibilities** of **developed countries** to climate change (IPCC 2014a: 127).

* I identify comments in this section as aligning with developed country positions as observed in IPCC approvals (WGII and WGIII for AR6) and UNFCCC Global Stocktake events (COP 27 and SB 58).

Table 7.2 *USA and UK authors in WGI's contribution to the AR4, as listed in the report.*

	Coordinating lead authors (22)	Lead authors (106)	Review editors (24)	SPM drafting authors (33)
US	7	25	4	12
UK	3	12	3	7
Total	22 (45%)	106 (35%)	24 (29%)	33 (38%)

assessment report and the writing team of the SPM (see Table 7.2). As Figure 7.2 identifies, these same countries continue to dominate the authorship of the SPMs in the AR5 and AR6, although Brazil, India and South Africa have emerged as important contributors.[7]

The government review also serves as an important constituent of a national delegation's preparations for the approval session (Zillman 2008: 33). The plenary approval proceedings and interventions by governments bury deep into the text of the SPM, the knowledge that underpins it and the IPCC rules by which it is compiled. Those that participate in approval proceedings with the objective to strengthen, weaken or 'improve' the text must arrive prepared with arguments supported by material contained within the assessment or on the grounds of the rules of procedure for compiling the assessment. This is no small task; there are three WG reports each with more than 1,000 pages of text that is condensed into a 30-page summary. The content of these documents has relevance for and bearing on work across government departments. Member governments resourced and invested in the IPCC process ensure that the appropriate expertise has been canvassed from within and outside of government to review this material and inform a national position on the text.

Those delegations arriving at the approval session without the support of a national review process are less well armed to suggest revisions that align the text with national needs and the international negotiating position. Without technical expertise to either inform government's preparations prior and/or within the national delegation, member governments cannot contribute to or take an informed position on technical issues, which confines a government's interventions to general comments. In an IPCC survey of national focal points, 31.6% of developing and economies in transition (EIT) country respondents did not carry out a government review of any of the TAR or AR4 products, compared to 12.5%

[7] Brazil, India and South Africa have held key roles in the bureau. For the AR6, South Africa and India co-chaired WGII and WGIII, respectively, and Brazil co-chaired the Task Force on National Greenhouse Gas Inventories (see Figure 7.2 for further details).

of developed countries surveyed, and the actual figure for developing countries is probably far lower than the 19 developing and EIT county respondents (IPCC 2009o).[8]

Once the drafting team receive the final collated government review comments, they have roughly two weeks to re-draft. This is done most intensively once they arrive at the approval venue, a few days before the session's opening, where co-chairs also prepare the team for the session. The process of selecting and crafting the chapter's key findings and conveying the overall message of the WG's assessment into this 30-page summary has its own tensions and divisions. At this stage, attachments to favourite sentences become apparent (Broome 2014), and disagreements arise over the presentation of the issue and the message it conveys (Schneider 2009: 166–68), energy that must be carefully channelled into the approval ahead. This is the moment for co-chairs and authors to let go of the text that they have spent weeks and months crafting so that it may pass into the hands of the waiting delegates as smoothly as possible.

7.3 The Politics of Approval

The approval session is the final destination of an SPM, bringing together those that oversee and author an assessment with those that approve how its key findings are presented to the world waiting beyond. This identifies the three main sets of actors or characters in the politics of the approval: the co-chairs, the delegates and the authors. Until this moment in the IPCC's practice of writing, the political activities of member governments and the scientific conventions of the authors that govern their respective contributions have remained largely separate. The practice of approval, however, brings scientific practices for constructing knowledge of climate change in direct contact with and subject to the political interests of member governments. Bureau members and authors combine their authority with the practices of plenary to limit governments' incursions into the text, and it is through the interplay of member governments, co-chairs and authors' ways of conducting their respective roles that the final document is written. The aim of this section is to explore the unique forms of authority and strategies that each set of actors have to structure the proceedings and influence this final component of writing climate change. To do this, I take each actor in turn, beginning with the co-chairs.

[8] Of the 19 developed countries and economies in transition that responded to the survey, six countries did not carry out a government review (IPCC 2009o). However, due to the low number of respondents, it is likely that the actual percentage of developing and EIT countries conducting a review of IPCC materials is lower than the 68.4% suggested by the survey.

7.3.1 The Co-chairs

The WG approval sessions are organised by the IPCC secretariat and, apart from when the COVID pandemic moved them online, are roughly a 5-day meeting hosted in various cities around the world on government invitation. The sessions are predominantly chaired by the WG co-chairs with support from the wider bureau, and historically, it has been the convention for the developing country co-chair to open the session and for the developed country co-chair to assume the majority of the session's chairing. The seating arrangements at these sessions are the same as for the regular plenary, with national delegations in alphabetical order followed by observer organisations and lead authors at the back of the hall, see Figure 7.3. On the podium or dais at the front of the room sit the WG co-chairs, the authors presenting the section's key messages and TSU and secretariat staff providing the technical and legal support for the process. For the AR6, the approvals of the WG reports were held virtually as in Figures 7.3 and 7.4, with the meeting scheduled over a two-week period.

Scholars frequently use metaphors of theatre to describe the conduct of intergovernmental meetings (Death 2011; Campbell et al. 2014; Craggs and Mahony 2014; Hughes and Vadrot 2019) and staging for the audience-dependent presentation of scientific knowledge and assessment processes (Hilgartner 2000; Gustafsson 2019;

Figure 7.3 The arrangement of the plenary approval session for WGIII's contribution to the AR5 in 2014. The text is projected at the front of the room, and the co-chairs, section authors, TSU and secretariat staff are seated on the podium. Delegates below are seated in alphabetical order with observer organisations behind. Photo by IISD/ENB reporting services: https://enb.iisd.org/climate/ipcc39/11apr.html.

Figure 7.4 A screenshot of the virtual approval of WGIII's contribution to the AR6. Photo by IISD/ENB: https://enb.iisd.org/56th-session-intergovernmental-panel-climate-change-ipcc-56-14th-session-working-group-III-4Apr2022.

Schenuit 2023). In terms of describing and analysing the practice of approval, choreography is useful for highlighting the preparation and planning undertaken by co-chairs, TSU and the wider bureau for these meetings.

Chairing is a central element of the approval practice. Interviews and discussions indicate that within the organisation, chairing is viewed as an art form and is an admired skill (interview 1.07.2010). While it is a skill largely acquired through observation, shared evaluation of past performances and practice (interview 5.08.2010), there are also institutional attempts to ease initiation into this role. Anticipation for the potential controversy of the SR1.5, for example, led the secretariat to host a training session with previous chairs (interview 26.02.2019). Co-chairs have also organised their own informal sessions with delegates to gain a government's perspective. This is critical and points to the importance of distinguishing between the role of chair, author and delegate in the practice of approval, as despite this being a shared undertaking, the distinct interests and activities of each of these actors give rise to different and even conflicting understanding and perceptions of the purpose and outcome of this final stage in the IPCC's practice of writing.

Scheduling is a second key element for the choreography of the meeting. The SPM is over 30 pages long, arranged into headline statements with paragraphs and figures that elaborate and support these, as well as signposts to the relevant sections of the underlying report. The co-chairs have to decide how to stage the presentation of these sections; this may follow chronological order, or a decision may be taken to front load potentially controversial elements to give delegates sufficient time to reach agreement (interview 26.02.2019). Alongside the timing of the content is selecting the 'right person' to chair the section. While developed country co-chairs continue to undertake the majority of chairing, no single individual can chair a meeting that will eventually run across day and night. Consideration needs to be given to where conflict lies, and which chair or bureau member may be best positioned to mediate it (interviews 5.08.2010; 9.11.2010; 26.02.2019). This carefully planned meeting choreography is captured in the podium document, which provides a shared script of the meeting for the bureau and TSU. Once underway, maintaining communication between the bureau, TSU and authors is critical and methods of whispering and passing bits of paper have now been replaced with slack (interviews 9.11.2010; 26.02.2019), a messaging app that allows for rapid sharing and collaboration.

With the backstage scripted and the front stage prepped, the approval session opens with the assigned chair introducing the SPM and detailing amendments made to the final draft in response to government comments. The timing of the release of this document is another carefully choreographed element (interview 26.02.2019). Releasing the final SPM too far ahead of the session gives governments time to bury deeper into the text and develop strong, well-informed positions. Released too late, and the start of the session can be mired by complaints over insufficient time to examine the revised text, which can induce ill-feeling that carries over into proceedings. With the SPM text projected overhead and the first section highlighted in yellow, attention shifts to the delegates waiting below. While chairs may have control over the preparation of the text and choreography of the meeting, once the session is open, the text moves into the hands of member governments.

The passing of the text from the chairs to delegates reveals the different perspectives. Co-chairs have spent months crafting every sentence and are acutely aware of authors investment in the document that is now projected on the screen. It is therefore unsurprising that co-chairs and authors share a view of success that sees the text pass through the approval as unscathed as possible. However, becoming possessive of the text and the activities of its passage can alienate the plenary participants, including other WG bureau members, as each awaits their turn in proceedings (interviews 26.07.2010; 9.11.2010). Therefore, effective chairing requires letting go of the text and using other resources and strategies to shape the meeting dynamics and their imprint on the SPM. Time, in the sense of setting and attempting to maintain the pace of the session is a key resource in this regard. An experienced chair knows that delegates want the opportunity to say their piece, may even relish a sense of urgency and plays on time accordingly (interviews 26.07.2010; 9.11.2010).

Figure 7.5 The state of progress on day three of the WGII approval session of the AR4. Photo by IISD/ENB reporting services: https://enb.iisd.org/climate/ipwg2/.

Experienced chairs speak about taking it slowly in the beginning and earning the trust of delegates by listening and responding to their concerns, then cashing in on this trust later once everyone is invested in the process and sensitised to the time constraints and joint endeavour of getting words, sentences and paragraphs approved.[9] Having only four to five days to approve the text, urgency is easily created, and it is not uncommon for discussion to stay circling the first paragraphs at the end of the second day or for difficult portions of the text to be pushed back until later in the proceedings. However, delaying the approval of contentious issues until later in the week can backfire. This happened in the case of the WGII plenary approval of the AR4, see Figure 7.5, where the all-night session that most approval plenaries experience ran over into an extra day and was generally regarded as ill managed and 'laborious' (Gutiérrez, Kulovesi and Muñoz 2007: 1), with some participants claiming it was 'one of the worst meetings they had ever attended'

[9] See Peterson's notes on a contact group he observed at WGI approval of the TAR in 2001 (Peterson 2006: 175–82) and Skodvin's observations and conclusions from the WGII approval session of a 1994 Special Report (Skodvin 2000a: 161–68).

(Gutiérrez et al. 2007: 14), signalling the pressure and expectations that chairing is subject to.[10]

7.3.2 The Delegates

While bureau members are elected for the assessment cycle and their continuation in this role is dependent on re-election, the national delegate remains as long as the government invests in IPCC participation, and they hold their post in the relevant government department. Since the approval of the FAR in 1990, and according to the participant lists available, on average 111 member governments have attended the approval of the final SPM of an assessment cycle, the synthesis report, with an average of 2.5 delegates per country. While the total number of delegates has increased since the FAR (see Figure 7.6), the total number of member governments has remained relatively steady, with the highest number (129) participating for the approval of the AR4, and only 42 countries have attended the approval of every synthesis report in Figure 7.6. Developing country participation has plateaued in IPCC meetings. According to the IPCC's own figures, on average 75 developing countries attended the four plenary meetings that took place between 2014 and 2016 compared to an average of 134 attending UNFCCC COPs during the same period (IPCC 2016b).

The average delegation size masks significant variation. Out of 177 countries that attended at least one approval session captured in Figure 7.6, the majority (104 countries or 59%) were represented by a single delegate. Figure 7.7 identifies the 31 countries with an average delegation size greater than 2, which is indicative of the small number of countries most active in the meeting. At least two delegates are required to participate in simultaneous contact groups and/or huddles and to sustain a presence during the all-night sessions in the final stages of the approval. One experienced delegate suggested, you don't need more than three or four, 'maybe five … as long as you have got a range of expertise in the team' (interview 26.07.2010). The larger delegations identified often include cross-departmental expertise as well as specialist knowledge in the assessment under approval from within and outside government, and some listed participants may not be directly contributing to the delegation. Bureau members are included in the delegation counts and nearly all countries in Figure 7.7 have or have had a bureau member in one of the six assessment cycles, which accounts for some countries listed.

Delegation size does not necessarily correlate with level of participation in the meeting.

[10] This was also highlighted by a WGII bureau member in his response to the IAC questionnaire, writing that 'the chairing of the entire WG2 plenary for the Fourth Assessment by just one individual – including a final mammoth 24-hour plus session – was not very effective' (IAC 2010b: 228).

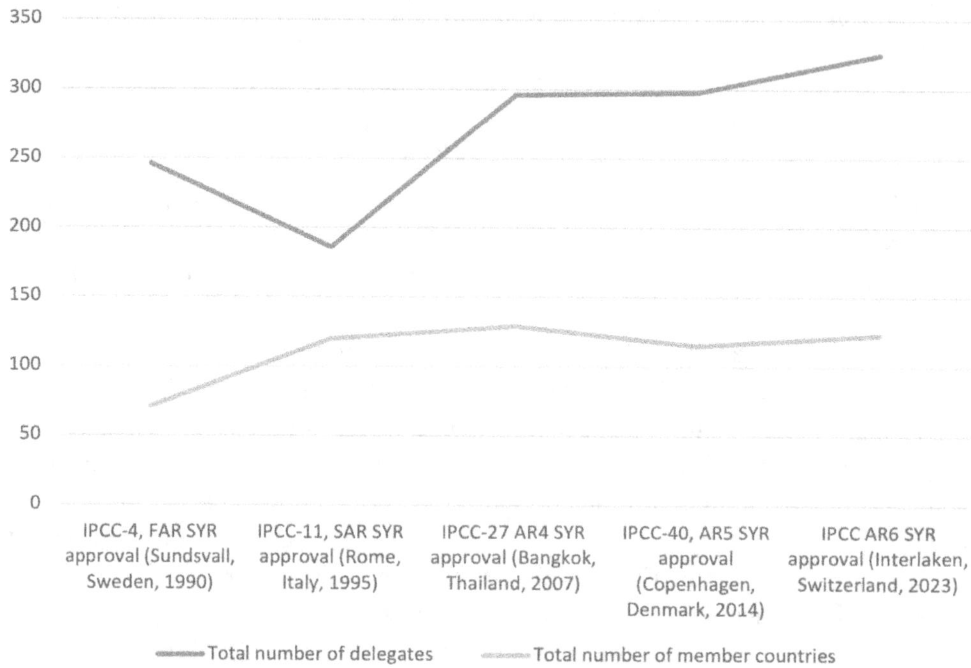

Figure 7.6 Number of government delegates and member governments attending the approval of the synthesis report for the FAR (IPCC 1990c), SAR (IPCC 1995), AR4 (IPCC 2007d), AR5 (IPCC 2014b) and AR6 (IPCC 2023), as recorded in the reports of the session.

Using the Earth Negotiations Bulletin (ENB) summaries as a measure of member government's engagement in the approval of WGI, II and III's contribution to the AR6, 29 member governments, plus the EU, are mentioned more than 20 times in total across these approval sessions (see Figure 7.8) (Bansard, Eni-ibukun and Davenport 2021; Eni-ibukun et al. 2022; Templeton et al. 2022). While Japan has on average the largest delegation, with 15 members, it intervened significantly less in the approval of the AR6 than India, whose delegation across synthesis approval sessions has averaged 2.8 and was 6 in the AR6 synthesis approval. Combined, Figures 7.7 and 7.8 provide a more precise sense of which member governments actively participate in the practice of approval. These figures suggest that out of the 100 or so member governments present at approval sessions, roughly 30 countries could be identified as core participants in approving the key findings of a report. Even across this core group, participation is uneven, with the EU and its member countries (24%), India (15%), Saudi Arabia (11%) and the US (9%) accounting for over half (59%) of the interventions recorded.

While the size of the delegation does not necessarily positively correlate with the number of interventions a member government makes, a delegation of two

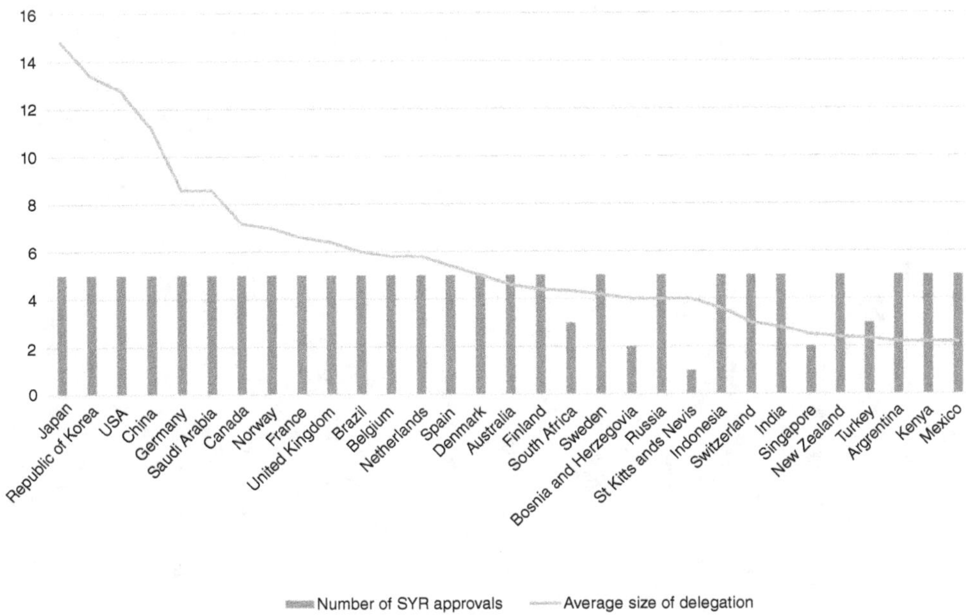

Figure 7.7 The 31 member governments with an average delegation size greater than 2 across the approval of the synthesis reports for the FAR (IPCC 1990c), SAR (IPCC 1995), AR4 (IPCC 2007d), AR5 (IPCC 2014b) and AR6 (IPCC 2023), as recorded in the reports of the session.

or more is essential for participation across contact groups and huddles. Contact groups and huddles are used to move polarised discussions out of the plenary and facilitate discussion of technical content between the authors and concerned parties. While contact groups are generally scheduled and chaired by a developed and developing country chair, huddles are chaired by a bureau member and may place on the side or even in the corridors. From observation of approval sessions, there is some coordination across UNFCCC negotiation blocks (De Pryck 2021). This enables smaller delegations with a shared position on climate change, such as AOSIS countries, to broaden their reach across the different sites of the approval, to echo and support interventions and to ensure that these shared interests are reflected in the emphasis and formulation of key findings.

As necessary as human resources are for active participation, delegation size and the number of interventions do not equate with symbolic power to shape the text. For this, it is necessary to identify the forms of authority that shape relations and distinguish the actors and delegations that have the greatest influence in and over the practice of approval. As described in Chapter 4, knowledge of the process, both in terms of IPCC processes and procedures and of the assessment's progression, are central forms of cultural capital, distinguishing delegates and ordering relations in the panel's practice of writing climate change. These forms of capital

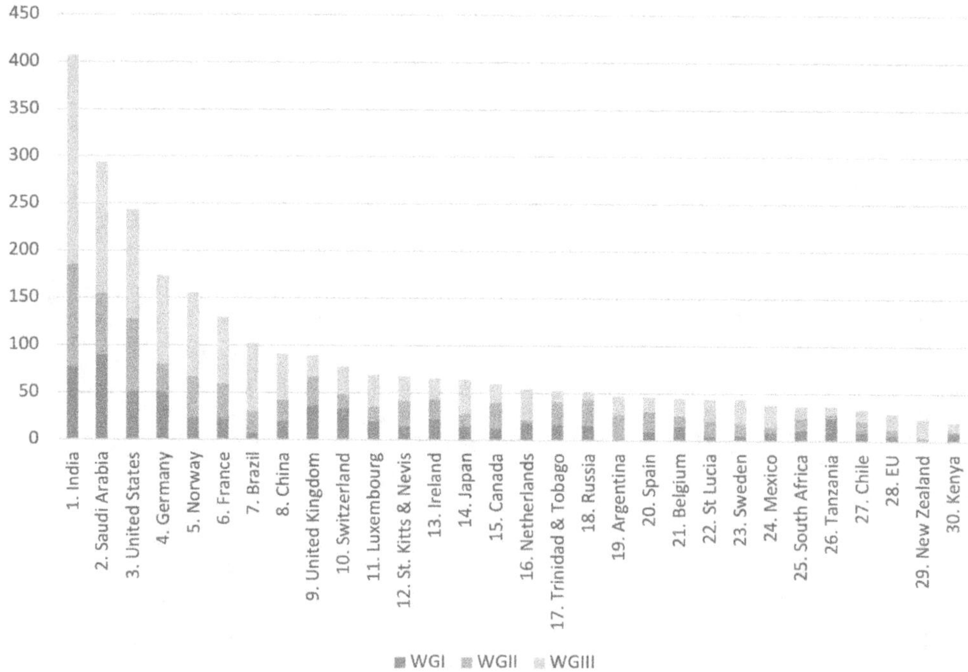

Figure 7.8 Graph of the 30-member governments mentioned more than 20 times, as recorded in ENB summaries for the approval of WGI (Bansard, Eni-ibukun and Davenport 2021) WGII (Eni-ibukun et al. 2022) and WGIII's (Templeton et al. 2022) contribution to the AR6.

are accumulated and embodied within long-standing delegates to the IPCC and are further enriched where a member government hosts the co-chair and TSU. This is most readily observed when we return to the approval in progress.

With the first paragraph projected on the screen, the delegates' role begins. This general introduction to the text is greeted by a wave of country flags. In depth knowledge is not required to intervene in a discussion on the scope of the introduction, which means nearly all delegations have an opinion on this constituent of the document or procedural issues to raise, and for some member governments this may be their main intervention in the meeting. Consequently, reaching a consensus on these three or four sentences can take up precious hours of the allotted time by running into a second morning or afternoon session. This highlights a number of important features of delegations, the properties of individual delegates and the tactics available to member governments to sculpt the SPM document and imprint their interests on the text.

Although nearly all central figures in the IPCC's establishment and/or early years have retired, there have been long-standing members of the panel that were influential in the formalisation of the IPCC's practice of writing. These delegates have sometimes fulfilled different actor roles, for example as bureau member or head of a TSU, or accompanied the national bureau member to meetings and served as the main point of contact between the government, chair and TSU. The time and financial commitment these member governments have made to the IPCC, such as through hosting meetings, chairing organisational working groups and funding a TSU is rewarded fourfold. First, through recognition of both the member government and delegate's support to the organisation (cultural capital); second, by the knowledge of the process that being a long-standing member and hosting a TSU enables (cultural capital); third, by the social connections that these activities foster (social capital); and fourth, the informal channels these relations create to additional know-how and perspectives on the process (cultural capital).

It is the combined cultural capital, in the form of knowledge of the process, and social capital, being known by and for, which enables some delegates to play a more active role in the proceedings, intervening more frequently with comments that are given greater consideration by the co-chairs and are influential over the thinking of other panel members. Due to their experience, these delegates may not be given detailed instructions from their government, which allows greater flexibility in how they play their role during proceedings. In some cases, the delegate's or delegation's participation is constructive to the process, enabling these actors to detect the direction in which particular disagreements are heading and intervene with suggestions that have successfully resolved similar issues in the past. At the same time, these delegates are also better able to phrase political interventions in knowledge of the process to assert their interests, which are more likely to be effective because of their symbolic power.[11]

Delegates and delegations become recognised and associated with these roles. The previous head of the British delegation, David Warrilow, see Figure 7.9, was well known amongst the panel, the secretariat, bureau and TSU members. As the British focal point, Warrilow attended plenary and bureau sessions from 1995 until his retirement in 2016. He also acted as a lead negotiator for the EU on the scientific issues covered in the UNFCCC. David was perceived by bureau members as a delegate that knows the science (and politics), and overtime he distinguished himself within the panel as a constructive member of the IPCC's practice of writing, regularly intervening in approval proceedings to offer advice on improving the wording, order and flow of the section under discussion. As a result, David's

[11] Adler-Nissen and Drieschova (2019: 543) make a similar observation, noting 'Negotiators who can achieve balanced wording and possess a certain institutional memory to recall previously "agreed language" that they can reapply to new circumstances, are the most likely to embed their preferred solutions in the final document.'

Figure 7.9 David Warrilow co-chairing a contact group at the 24th plenary session of the IPCC in Montreal, 2005. Photo by IISD/ENB: https://enb.iisd.org/climate/ipcc24/28september.html.

opinion was sought on all matters of IPCC business inside and outside of plenary sessions. He often headed task groups to gather background and opinion to inform panel decision-making, was commonly requested to chair contact groups during plenary and WG approval sessions and was often at the front of efforts to broker deals between dissenting parties.[12]

In other instances, a delegate's notoriety can signal their obstruction to the process. The Saudi Arabian delegation is comprised of a highly skilled team of delegates. In early assessments, the approval delegation was generally headed by Dr Mohammad Al Sabban. Mohammad Al Sabban, see Figure 7.10, was also the chief negotiator to the UNFCCC process from 1990 to 2012 and senior economic advisor to the Minister of Petroleum and Mineral Resources (now the Ministry of Environment) from 1997. He also distinguished himself as a member of the panel, albeit through a different mode of engagement than David Warrilow. The Saudi Arabian approach, led by Mohammed Al Sabban, was more commonly associated with hindering the approval proceedings. Regarded in the UNFCCC negotiating

[12] This reveals the importance of consistency in delegates and delegation over time. When these delegates retire, some of the symbolic power that they have accumulated is attached to the member government, which ensures the new delegate has a place in the existing order of relations they step into. However, the capital has to be carefully passed over and cultivated in and for the replacement so that the conduits to knowledge that have been created are not lost and are maintained and built on by the new delegate.

process as one of the key players and usually cast in the role of villain (Depledge 2008; Harris 2009), he gained notoriety in the IPCC for his performance during the approval session of WGI's contribution to the SAR in Madrid, in 1995 (Houghton 2008). At this approval session, notes were passed between the Saudi Delegation and the head of the Global Climate Coalition, Don Pearlman, with repeated objections that delayed the approval of the text (Leggett 1999: 224–30; Schneider 2009). This reveals that while time is a tool that the co-chairs attempt to command, it is also an instrument for delegates to play.

Delegations can attempt to delay proceedings by continually intervening, by raising issues with the text and by re-opening previously approved sections (Skodvin 2000a: 162–65). This is a tactic that the Saudi delegation has been associated with both in IPCC and UNFCCC proceedings (Depledge 2008). In the approval of WGI's contribution to the SAR, the Saudi Arabian delegation made life difficult for the chair and authors by repeatedly objecting to text (Leggett 1999; Houghton 2008; Schneider 2009). They also missed the huddle where delegates worked closely with the authors to craft agreeable language (Chemnick 2018). This enables a delegation to re-open debate when the proposed text is returned to the plenary on the basis they did not participate in discussions and further delay the progression of the meeting. However, tactics like these have to be used sparingly or they result in lasting bad feeling that can impact a member government's symbolic power and hamper their ability to manoeuvre effectively.

As US government wikileaks reveal, there was some recognition within the Saudi Arabian government that Dr Al Sabban may have lost his capacity to read the changing mood that took place with the negotiation of the Copenhagen Accord at COP 14 in 2009, and to adapt the countries position accordingly (Guardian 2010d).[13] During the AR6, there was a change in the membership of the Saudi delegation, with the all-male team replaced by a younger, female-strong delegation led by Dr Malak Al Nory (see Figure 7.10). In light of the comments captured in the Wikileak (see footnote 14), this change of guard appears as a well-orchestrated move to regain and retain the national position as a symbolically powerful and

[13] At the start of COP 14, and in response to the leaked Climategate emails, Al Sabban suggested the emails called in to question the human cause of climate change and that the incident would have a 'huge impact' on the negotiations and countries willingness to cut emissions (Black 2009). After the negotiations and informed by Sabban's analysis, the Saudi delegation did not believe the Copenhagen Accord would attract significant support (Guardian 2010d). As a result, and as the WikiLeaks identifies, 'The Minister's office was unpleasantly surprised by mid-January, when it was clear that a number of countries had already associated themselves with the accord' and there was a sense articulated by Assistant Petroleum Minister Prince Abdulaziz bin Salman '*that Saudi Arabia had missed a real opportunity to submit "something clever", like India or China, that was not legally binding but indicated some goodwill towards the process without compromising key economic interests.* The Prince intimated to EconCouns that Al-Sabban would not long retain his position, and said the challenge for Saudi Arabia was to find a way to "climb down" from its negotiating position' (Guardian 2010d).

effective writer in and of the IPCC's practice of approval, which is vital to achieving their overall negotiating aims in the UNFCCC.

As this account of a symbolically powerful member government in the IPCC's practice of approval makes apparent, the motivation for accumulating symbolic power and the role adopted in proceedings is in large part driven by the national position on climate change as negotiated within the UNFCCC, although that is not the only motivation. This results in a degree of regularity and even predictability to the approval proceedings, which extends beyond the routines and conventions of doing an IPCC approval plenary into the timing and content of delegates' interventions. The Saudi Arabian delegation provides the clearest illustration of this, but it is not alone in this role. Since the approval of the FAR in 1990, Saudi Arabia's interventions have focused on the confidence levels assigned to the scientific findings and preventing carbon dioxide from being distinguished from other greenhouse gases (Leggett 1999: 17), which initiates intervention every time relevant terms appear in the text. The content of Saudi Arabian interventions continued to question the certainty of scientific claims in the AR4, with four interventions recorded in the ENB summary of the WGI and WGII approval session of Saudi Arabia objecting to the certainty language employed (Gutiérrez, Kulovesi and Muñoz 2007; Gutiérrez, Muñoz and Johnson 2007). In one case, China and Saudi Arabia proposed reducing or qualifying the probability that anthropogenic greenhouse gas increase has *very likely* caused most of the observed increase in global temperature by removing the adverb 'very' or adding the term 'increasingly' *very likely* (Gutiérrez, Muñoz and Johnson 2007: 5).

In the AR6 cycle, Saudi Arabia moved its focus on to dampening the emphasis on emission reductions and fossil fuel phase out. In the approval of WGIII's contribution to the AR6, there are seven recorded instances of Saudi Arabia intervening to add the word 'unabated' to sentences referencing CO_2 and GHG emissions reductions, in one instance calling for retaining language 'on avoiding unabated fossil fuel emissions rather than "displacing" fossil fuels' (Templeton, et al. 2022: 19). This indicates that while a country may accept that some battles are lost, for those parties with deep interests, be those economic or physical survival, the energy of the struggle remains, and attention is shifted to shaping new objects and concepts that have the potential to significantly shape the negotiating process. Saudi Arabia is an easy target for analysis in this regard because its interests in fossil fuels are deep, but so are those other countries, which can leave the objections and the cultural capital expenditure to the Saudi Arabian delegation.

While the UK and Saudi Arabia have been active participants since the IPCC's establishment, some actors have deepened their involvement in the panel and participation in the practice of approval over time. In the written accounts of the FAR and SAR, there are few references to interventions by

China. However, by the AR4, China sends one of the largest delegations with an average of 15 delegates and begins to play a core role in approval proceedings (Gutiérrez, Kulovesi and Muñoz 2007; Gutiérrez, Muñoz and Johnson 2007; Schneider 2009: 180–97). In the AR6 and according to ENB recorded interventions, China was the eighth most frequent intervener, making up 3% of total interventions during the WG approval sessions (see Figure 7.8). Caribbean member states also emerged as core participants during the approval of the AR6. Combined Saint Lucia, Saint Kitts and Nevis and Trinidad and Tobago are mentioned 165 times across the ENB summaries for the WG approval sessions, which is equal to 6% of the total interventions (see Figure 7.8). St Kitts and Nevis did not appear on a synthesis report approval participant list until the AR6, when it arrived with a delegation of four (with three registered participants from Climate Analytics[14]). More research is required to understand a country's changing level of involvement. However, as context, it is important to note Caribbean member states activism in having the 2°C temperature goal re-evaluated from COP15 at Copenhagen, on the basis that it undermined the survival of their communities (Tschakert 2015: 2). These calls initiated a process of structured expert dialogues to assess the adequacy of the long-term goal and eventually led to the invitation for a special report on the impacts of 1.5 in the Paris Agreement (Tschakert 2015). In the approval sessions, these states frequently intervened to support the authors and on issues related to the 1.5 temperature goal, impacts, emissions reductions, barriers to adaptation, loss and damage and urgency (Bansard, Eni-ibukun and Davenport 2021; Eni-ibukun et al. 2022; Templeton et al. 2022).

However, it is not only state interests that explain member governments changing participation over assessments. It can, for example, be the result of individual delegates investing themselves in the process and gaining confidence and ease in its navigation, particularly as their awareness of the potential for IPCC products to impact climate negotiations increases over time and is conveyed back to the government. Having someone within government elected to the bureau can serve to increase national interest and investment. This can be particularly important in the case of developing countries, who may use the additional funding to socialise another member of the government department into proceedings and thereby strengthen the government's capacity and expertise (interview on 29/03/2023).

Turning from the order of relations in the practice of approval to their effects on the text, in most instances, the contents of government interventions result in the substitution of words or re-ordering of sentences, which may even lead to improved clarity and communication of complex science. Of course, as the

[14] Climate Analytics is a global climate science and policy institute.

Figure 7.10 Left: The Saudi delegation led by Mohammad Al Sabban at climate change talks in Bonn 2010. Photo by IISD/ENB: https://enb.iisd.org/climate/ccwg11/ Right: Members of the Saudi Delegation headed by Malak Al Nory (left) during the virtual approval of WGI's contribution to the AR6, August 2021. Photo by IISD/ENB: https://enb.iisd.org/sites/default/files/2021-08/malak_al_nory_.jpg.

examples above indicate, adding additional terms or words each time a particular concept appears can also make the meaning of a sentence less clear, such as adding 'unabated' or 'other greenhouse gases' in front of every appearance of fossil fuels, carbon dioxide and emissions in the text. Requests by governments for clarity or to include additional findings and sentences from the underlying report can also greatly increase the length of the assessment, which on average increased by 17–53% in the AR4 and AR5 (Mach et al. 2016). WGIII's SPM for the AR6 grew by two thirds through the approval process from 31 pages to 53 (IPCC 2022), making it the largest ever summary document. Size matters in the communication of climate change; succinct key messages and powerful visuals facilitate travel; and it appears that in this instance some member governments may have sought to impede the travel of climate mitigation knowledge.

Despite the fact that the SPM always increases in length, deletion is another strategy of delegates in the practice of approval. Continuous intervention and objection – particularly once the pressure of time is bearing down on the proceedings – can succeed in getting sentences, boxes, figures and, in some instances, entire sections removed from the document. Even when time is allotted, if text has the potential to influence UNFCCC, negotiations it may be insufficient. This brings us back to country categorisation in the AR5. Zooming in on member government's comments on the final draft in box 7.1, it became apparent that while the Annex I, or developed countries, sought to have country categorisation re-evaluated along the lines of income, some developing countries with growing economies and GHG emissions sought to maintain the differentiation of the Kyoto Protocol. Although the contact group met over the course of three days, it failed to reach agreement, which led to the deletion of four figures and all relevant paragraphs from the final report (Gutiérrez et al. 2014: 8). The countries opposed to income categories cited their concerns that 'policymakers would draw on the SPM for the UNFCCC negotiations on a new climate agreement' (Gutiérrez et al. 2014, 8). This serves to highlight that there are some issues on which parties are so divided and the relevant objects of this division so heavily weighted by their potential effect on the negotiating process, even if a chair deploys all strategies available to them, ultimately the need to produce an approved text within the allotted time, or within a reasonable time beyond that, will result in deletion.

7.3.3 The Authors

This brings us to the authors. What strategies are available for authors to prevent the incursion of member governments into the key findings of the assessment? For authors, observing the proceedings from the back of the hall, this plenary-specific way of approving the text is, for most, a new experience that has been described as 'exceptionally frustrating' (Stavins 2014), slow, awkward and time-consuming (IAC 2010b: 38, 84, 112). Government interventions are often regarded as political, time wasting, and delegates enjoying the sound of their own voice. However, authors are also susceptible to being swept up in the unfolding theatre, with dramatic accounts of scientists storming out, refusing to alter the text, and more mundane anecdotes of keeping themselves amused by taking bets on the length of time between Saudi interventions (Schneider 2009: 138).[15] To some extent this reflects the role of authors in the proceedings and the intrusion of government delegates into the authors' sphere of influence over the science and key findings of the assessment in the SPM, which at times only leaves room for symbolic gestures.

Author awareness of the significance of this stage in the IPCC's practice of writing has grown over assessment cycles through author's published accounts (Schneider 2009; Broome 2014; Stavins 2014) and the sensitivity towards the political context instilled by the bureau through the drafting cycle. This anticipation of the struggle results in careful selection of key messages and preparation to defend them (O'Reilly 2022: 167). However, just as the chairs must let the text, and to some extent the proceedings, pass into the hands of the delegates, so too must the authors. Attempts to maintain control over the wording or refusing to see the knowledge from the perspective of governments as policy prescriptive and/or culturally situated can and has hampered the approval of the SPM, in one infamous instance even necessitating an additional session.[16]

[15] Although accounts differ (see Schneider, Chapter 6), in the approval of WGII's contribution to the AR4, a lead author was reported to have walked out of the meeting hall after the scientific certainty of a statement was lowered as a result of Chinese and Saudi objections (Eilperin 2007; Mason 2007; Vergano and O'Driscoll 2007).

[16] This happened in the case of WGIII's contribution to the SAR because of content referring to the statistical value of a human life, which was valued higher in developed countries, content that developing countries were unable to accept (see Section 4.6 and footnote 20).

The role of lead authors is to ensure that any suggested revisions to the text are consistent with the content of the underlying chapter and the literature informing it. Thus, when a new paragraph of the SPM is opened for approval, the authors responsible for that section take their place on the podium alongside the chair (see Figure 7.4), provide a short overview of revisions made and identify the evidence base in the underlying report. In opening the section for comments, the chair will remind delegates to keep their interventions brief and to offer concrete suggestions. Delegates' interventions identify concerns with the text and request further explanation and clarification. The majority of the ensuing discussions centre on clarifying the terms and concepts employed by the authors and translating them into a language that is comprehensible to the SPM's audience. As time passes and the next government is identified on the list, the chair persistently presses for concrete proposals. It is the role of the authors to indicate whether the proposal fits with or distorts the meaning of the chapter content and the literature that underpins it. The author's authority in these proceedings rests upon their in-depth knowledge of the subject area and their capacity to rule whether proposed changes are consistent with the content of the full report. However, this scientific authority does not operate unchallenged. The right of authors to rule over the text becomes a constituent of the struggle within and between delegates seeking to uphold their interests through the practice of approval.

In cases where disagreements over text appear unresolvable, a contact group or huddle is formed, depending on the type of issue and the number of interested parties. This practice, which outside of translator's working hours proceeds in English, enables governments objecting to a particular word, sentence or section to work alongside the authors in a more intimate setting and broker text that can then be taken back to plenary for approval, a process that can be completed in a matter of minutes or run into days. During these sessions delegates can question the authors on the meaning of scientific terms and phrases, which necessitates authors translate technical understanding into a language communicable to a wider audience. Many of the delegates present will be UNFCCC negotiators, and this process enables them to establish the political content of a concept and its potential to bear on the negotiating process (interview 5.10.2010). As one author notes, it is within these contact groups 'behind closed doors in small groups', that motivations for delegate's interventions were made explicit and 'representatives worked to suppress text that might jeopardize their negotiating stances in international negotiations' (Stavins 2014).[17] Authors may step outside of the room to discuss and redraft between themselves, refer back to the underlying literature and

[17] Unlike in the plenary, country names are not attributed in ENB reporting of contact group and huddle discussions in the IPCC.

contact members of the wider chapter team before they offer or agree a suggestion. In the most controversial sections, new proposals are greeted with an additional round of comments, with delegates sending a photo of the wording back to government (Broome 2014: 12) or phoning a minister for further instructions (interview 4.10.2010). Through this back and forth, approvable language is crafted that neutralises political content, renders it opaque or removes it altogether.[18]

As highlighted, however, there is no guarantee that this text will be accepted by the plenary, and while in some cases a few minor adjustments are all that is required, on other occasions authors have found themselves embroiled in the same disagreements that initiated the contact group in the first place. As described in relation to Saudi Arabian interventions during the SAR, delegations may use their absence in the discussion as a reason to reject the revisions.[19] In the face of these tactics, authors have little at their disposal to constrain government behaviour and must contain their frustration, as author outbursts are not well received by delegates. When Mohammad Al-Saban raised a series of objections to text that had been agreed upon in the contact group, the lead author, Ben Santer, lost his temper and responded that the issues could have been discussed in the contact group had a member of the delegation been present. Al-Sabban replied that it was his job, as author, 'to serve the governments of the world, not to have an independent opinion' (Chemnick 2018). In this instance, Saudi Arabia became an isolated figure and faced with being the only country identified in a footnote, withdrew its objection to the sentence.

There are instances, however, where issues are so politically charged that no common ground can be found, as observed in returning to the country grouping example in the AR5. Despite three days to work together in a contact group, all relevant figures and content on categorising countries in relation to income and GHG emissions were deleted from the SPM, including any reference to relevant content in the underlying assessment (Gutiérrez et al. 2014). In situations like this, and as tensions mount, both authors and delegates may resort to threats (Broome 2020), such as the threat of walking out, resigning from the author list, recording dissent in a footnote or making the content public. This is what happened as a result of lost content on country groupings and international cooperation in the AR5. The authors involved published all deleted figures, an account of the event and its implications for the future of the IPCC in a commentary in the journal *Science* (Edenhofer and Minx 2014; Victor, Gerlagh, Baiocchi 2014; Wible 2014). The drafting author of deleted content on international cooperation wrote his grievance in a letter to the bureau, which he later published on his personal blog along with

[18] For accounts of this in the literature, see Petersen 2006, appendix: 113–17; Kouw and Petersen 2018.

[19] For smaller delegations this is a genuine problem, as oftentimes there are several contact groups running simultaneously throughout the approval session.

the deleted section (Stavins 2014). In the end, these are largely symbolic actions, more impactful on the scientific field than member governments, who excluded this content as the basis of collective action in deleting it from the SPM. However, the example serves to highlight that to understand what the practice of writing produces in an assessment – what is or is not contained in the final SPM – attention needs to be given to the choreography of the meeting, the dynamic between chairs, delegates and authors and the presence of particular characters. More critically, the analysis must be situated within the negotiating context of the UNFCCC and governments positions on climate change within the negotiations, as this generates the forces structuring the practice of approval and the content that results.

7.4 Summing Up

I opened this book with the story of four countries – the US, Saudi Arabia, Russia and Kuwait – refusing to welcome the IPCC Special Report on 1.5 at COP 24 in 2018. As the account of authors publishing deleted content in *Science* also indicates, the practice of approval does not contain the struggles it initiates. These forces originate from and overflow back into the broader field of global climate activity, where the interests in climate science and politics are constituted and which ultimately drive actor roles and the strategies they deploy in the practice of approval. Not all participants are happy with the final product and its potential consequences, hence Saudi Arabia's attempt to distance itself from and dampen the reception of the SR1.5 and the authors' decision to publicise deleted content. Recounting these stories as part of the politics of approval identifies the IPCC as a central site in climate agreement-making. As member governments have grasped the impact of IPCC knowledge on UNFCCC negotiations through its provision of objects and methods for determining the distribution of collective responsibilities, they have deepened their involvement in the practice of writing and brought the negotiations into the approval. The chapter documents the forms of authority and strategies that co-chairs and authors have to channel and contain these forces through, for example, the choreography of the meeting, the knowledge of authors and cultivating an attitude of openness to its collective re-writing. However, as evidenced through delegations own delaying tactics, all forms of authority and strategies become available sources and resources in member governments' attempts to re-write the meaning of climate change.

It is the interplay between these different actor roles and strategies, as situated within the broader context of climate politics, that are constitutive of the politics of approval and its imprint on the SPM. As in all aspects of the IPCC's practice of writing, the capacity to participate in the approval session, adopt these roles and deploy these strategies to shape the text is not equally distributed and is ultimately dependent on a member government's interests and economic resources to invest. Governments invested in the IPCC and its practice of writing climate change undertake an extensive review of IPCC materials, particularly the SPM, which prompts the most discussion amongst drafting teams and ensuing revisions in the re-drafted policy document. It is the same governments that arrive at the plenary prepared through the expertise enlisted for the review and as contained within the delegation. This relationship is strongest when a country chairs a WG and hosts the TSU, which is a significant economic investment in the IPCC process. These countries tend to have the highest number of authors in the SPM writing team and the greatest knowledge of the assessment process in practice, which enables informed position-taking on the text. While recognition of these governments' contributions can complement these forms of symbolic power – making authors and chairs more amenable to their comments – it can also constrain a government's capacity to diverge too far from the science that national authors wrote.

Other delegations adopt roles that do not impose restraints on their capacity to bend the practice of writing to their interests, which brings Saudi Arabia into view. However, it is in documenting the deepened involvement of developing countries in the practice of approval that is revealing of both how order imprints through the practice of writing and how the order of relations within the IPCC and the world beyond – the global distribution of economic, cultural and social resources – is changing. This is documented in this chapter through the emergence of new countries as contributing to the drafting of the SPM and in changing relations of participation in the approval. It is also evident in the struggle over content that attempted to regroup countries based on income and GHG emissions, which caused some of the greatest struggle and deletion in an SPM. This highlights the extent to which social order matters in the IPCC – both as a reflection of the broader distribution of global resources and as a distribution of power to write the future order, including the basis by which resources are valued and distributed – in the naming of climate change. While the economic stakes of climate change define the interests of all actors in the IPCC, for some, climate change is ultimately and urgently an issue of survival. Those countries that do not have the capacity to significantly shape the content through authorship of the scientific assessment, such as Caribbean member states, must leave their mark on the writing of climate change in the politics of approval.

8

Concluding on the Meaning and Implications of Writing Climate Change

I started in search of those with the power to name climate change. I wanted to know why governments were not tackling this issue in a meaningful way, why greenhouse gas emissions continued to increase and why everything appeared to stay the same, while the climate system changed. The IPCC was the site to explore this, and I found an answer in the practice of writing. It is not the answer I expected. And at first, all I saw was scientific and political activity channelled into an exercise of building an international assessment practice that served to maintain the existing order. I wondered at the true utility of this and where climate change was in these activities, as they appeared to continue as they were before and after its discovery and with every scientific alarm that followed. With time this sense faded, because when you watch something for long enough, you see change and that is what I have come to see – a re-making of the order of relations in and through the IPCC and its practice of writing climate change, but not exactly as it was before. It is also a consequence of expanding my site of observation, slowly moving outward from the IPCC to the field of climate activity it is situated within and grasping the effect of this situation – the forces it generates – on how the IPCC practices its assessment and names climate change and the impact this has on collective agreement-making.

It is the analytical approach of the book that enables the IPCC and its assessment reports to be understood as sites and products of agreement making (Hughes et al. 2021; Hughes and Vadrot 2023). There are two dimensions to the book's analytical approach. The first is the capacity to situate the IPCC within and as a component of the broader struggle over climate change and the field of political activity this generates. In Chapter 3, I described the struggle to name climate change as ultimately a struggle over the distribution of social, political and economic resources or order and the values that underpin these arrangements. I identified the IPCC – the main knowledge provider – as the central site in global attempts to determine the meaning of this problem (Hughes 2015). To fulfil its mandated task and to name climate change, the IPCC has developed and institutionalised a practice for producing assessment reports, what I refer to as the practice of writing. The second analytical dimension is the framework of actors, activities and forms of authority outlined in Chapter 4. This approach makes it possible to identify the social order of the IPCC's practice of writing, its relationship to the broader pattern and distribution of economic resources, how this order imprints on the writing of climate change, and how it is challenged and changes over time.

8.1 The Model of Science in Politics

When I began this research, I did not understand how centrally science is situated within politics and politics within science. Chapter 2 records my journey through models of science in environmental politics. The epistemic community model proposes that scientific knowledge comes before political action. In some respects, this reflects the emergence of climate change and other environmental issues at the time. Peter Haas was looking at emerging environmental problems and conceptualised the role that communities of transnational scientists had in constructing these issues that informed and shaped political action and the formation of new institutions (Haas 1989, 1990). The epistemic community model was influential in documenting the emergence of climate change on the political agenda. However, as scholars observed the global environmental treaty-making process more closely, such as Karen Litfin's study of the Montreal Protocol (Litfin 1994), it became apparent that the underlying assumptions that science informs politics and that this influence is unidirectional did not hold.

Litfin's work revealed that often it was not the scientists that were communicating the scientific knowledge on the discovery and extent of the issue, but actors that emerged through the treaty-making process – knowledge brokers – actors that proved adept at framing the science for policymakers. Litfin's study identified knowledge as something of a public good, available to all actors in the negotiating process to incorporate and deploy in their discursive strategies, rather than the preserve of its scientific producers, as suggested in the epistemic community model. This unravelling of the underlying assumptions continued as research turned from the physical and biological scientists that were central in initially identifying the problem to the fields of knowledge assessing the social and economic consequences of climate change, which identified struggle rather than consensus between disciplinary ways of knowing a problem (Bernstein 2001).

The epistemic community model remains a valid starting point for research interested in the role of science and communities of scientists in treaty formation. It sensitises the researcher to the balance struck between science and politics and government attempts to control the impacts of knowledge in the institutionalisation of global advice. However, once a practice is in formation for the purpose of approving a knowledge base for collective action, the epistemic community does not offer an accurate depiction of the relationship between science and politics. I have observed and documented three sides to this relationship in the IPCC's practice of writing. On one side, there are the scientific communities invested in the production of climate knowledges from diverse disciplinary perspectives. The scientific conventions and measures of authority that structure the production of knowledge within these fields underpin the practice for writing climate change and order author relationships in its writing (Chapter 6). As the internationally recognised site *for* writing climate change, the IPCC and the field of climate politics that it is situated within also became forces in climate knowledge production (Hughes and Paterson 2017).

On the other side, there is the political activity orientated around responding to climate change, which again has its distinct forms of interests and measures of power that structure and orientate action. The IPCC is a central object or force within climate politics because its assessments provide the knowledge base for negotiated action as well as methodologies for reporting on action. As a result, member governments of the panel have increased their power in and over the IPCC's practice of writing (Chapter 4). This is evidenced in the struggle over the election of the bureau and the approval of the outline (Chapter 5), as governments attempt to maintain some control over how climate change is written through the next assessment.

The third side of this relationship are the places where, by necessity or through design, the science and politics of climate change are brought together, as in the IPCC's practice of approval in Chapter 7. At these sites and for specific purposes, climate knowledge producers and government delegates negotiating collective climate action and/or informing the national position on negotiated action are brought together for a set task, such as approving the wording of an intergovernmental assessment's key findings. The practice of approval is not the only site within the IPCC or the broader field of global climate activity where science and politics are brought together by design to exchange and/or craft a particular outcome, such as an SPM, a workshop report or standardised methods for national reporting. Sites for these exchanges are organised within UNFCCC meeting sites and coordinated with the publication of an assessment or special report through UNFCCC/SBSTA mandated, IPCC organised and author-led events, including expert dialogues and meetings, facilitated exchanges, workshops, side events and more impromptu at booths.

The Global Stocktake (GST) further institutionalised the design of sites and activities to bring climate science and politics together within the UNFCCC. The GST provides for a periodic stocktake of the implementation of the 2015 Paris Agreement 'to assess the collective progress towards achieving the purpose of this Agreement and its long-term goals' (UNFCCC 2015, Art 14). The outcome of this five-yearly process is designed to inform parties 'in updating and enhancing' nationally determined contributions and collective efforts (UNFCCC 2015, Art 14), and ultimately for enhancing collective ambition in addressing climate change. The latest IPCC assessment reports are identified as a source of input as the 'best available science' for the GST, and during the technical phase a series of world cafés, roundtables, exchanges and poster exhibitions were thoughtfully crafted to facilitate exchange between authors and bureau members of the AR6 and government delegates (many of whom had approved its key findings), to establish the shared knowledge base from which collective progress could be measured.

This designation of the IPCC within the Paris Agreement is likely to have further increased the pressure on the practice of approval in the AR6 and beyond – as any object within the SPM has the potential to travel into and become a force to measure collective implementation. What I take from my observations across these different sites is that (climate) science and (climate) politics are always producing climate change as an object of knowledge and action and are – as forms of knowledge and action – continually being produced through the necessity to address climate change. In this way of thinking, science is not separate from or informing action; it is a central and constituent part of collective action or agreement-making on climate change.

8.2 Actors and the Forms of Authority That Matter

It is the interface between science and politics that has led scholars to study the IPCC as a boundary or hybrid organisation. As I have described in the different sides of this relationship, the origins of organisational practices and sources of authority in the IPCC are amalgamations drawn from both science and politics (Guston 2001; Miller 2004). This is evident in the practice of approval, where delegates learn the intricate details of the science of the underlying report for authoritative reasons to alter the text, and where authors negotiate this re-wording to avoid and accommodate political sensitivities. However, as I got closer to the IPCC, attempting to perceive it solely through its scientific and political content constrained the actors and activities that could be analysed as constitutive of the practice of writing. One of the main motivations of the study became to describe all the different actors that make up the IPCC, the activities they undertake and the forms of authority this gives them in and over the assessment. The analytical framework of the book, actors, activities and forms of authority, makes it possible to document the historical emergence of a social order within the IPCC's practice of writing climate change and to identify the distinct properties that are valued within the organisation.

The description of the IPCC and the order of relations in Chapter 4 reveal that while scientific and political authority remain central determinants of the culture and thereby the social order within the organisation and the conduct of the assessment, they are not the only forms of authority that matter (Hughes 2023). Those that led the establishment of the organisation had to find a way to fulfil its mandate – to produce assessments on the science, impacts and response measures to address climate change. This was not only a significant scientific undertaking; it was a huge administrative task. Realising a global assessment is dependent on the everyday seemingly mundane activities of scheduling and organising meetings, compiling and editing drafts, and harmonising and preparing final versions. The extent of these activities has grown as the fields of climate knowledge, the political demands and external scrutiny have expanded with each assessment cycle. This has required actors within the TSUs to codify authorship of the assessment and instil its importance to ensure the accuracy and rigour of the final report. This reveals the significance of the TSUs as a unit within the IPCC. The organisation depends on the TSU for realising an authoritative assessment, and actors within the unit, through their proximity to the emerging report, have the most in-depth knowledge of its progression, giving TSU actors unique forms of authority in and over the IPCC's practice of writing.

The TSU's forms of authority are also sources of capital for actors in the organisation that have close social relations to them and thereby access to and conduits for their knowledge on the assessment in practice to flow. These forms of capital are most readily available to the developed country co-chairs and governments that host these units. For the co-chair working alongside, this ensures their vision and leadership for the assessment materialises in the final product; for the hosting government, it equals the symbolic power to speak, be heard and to effect the decision-making of the panel, the writing of climate change and the rules by which climate change will be written (of which there is no greater power in the IPCC). The UK and US have hosted TSUs for 5 out of 6 assessment cycles, ensuring it is the culturally valued properties of these countries that govern the order of relations in the panel, bureau and the practice of writing.

TSUs return power to the powerful. However, the technical and administrative authority they hold means that they have potential to act simultaneously as upholders and re-makers of the order of the IPCC's practice of writing. Their role in upholding the scientific authority of the assessment is most visible during author selection, when the TSU applies scientific conventions for measuring a candidate's research impact and productivity. As Chapter 6 indicates, this can produce author lists that, if left unchallenged, reflect developed countries dominant position in the global knowledge economy. When the organisation establishes diversity criteria to ensure geographically and gender diverse authorship, it is the TSUs that must find a practical way to identify the expertise that meets these criteria and can fulfil the government-approved outline during author selection (Standring and Lidskog 2021). Equally, when a more diverse authorship is appointed, it is the bureau and TSUs that have the capacity to ensure this translates into greater participation in the authorship of the report and broader perspectives on climate change written into the final product. Through organising diversity training, surveying and collating author views and establishing themselves as a point of contact to identify exclusions and harassments (IPCC 2019), the TSUs can play a key role in challenging the pervasive and reproductive character of scientific conventions and measures of authority in the IPCC's authorship of climate change. Either way, TSUs are order makers in the IPCC's practice of writing.

The book's account of the emergence of the TSUs (Chapter 4) and their capacity to structure the order of relations in the authorship of the assessment (Chapter 6) reveals that the forms of authority operating in the IPCC and structuring the order of relations in the organisation and its practice of writing continue to evolve with the forces and pressures exerted on the IPCC by its centrality in and to climate politics. This is also apparent in the emergence of the secretariat's role in and authority over managing IPCC media relations (Section 4.5). It is indicative of how an organisation, like the IPCC, has to continue to evolve in response to the pressures and forces that are generated by its situation within a field that its products shape. It also reveals that even in an organisation identified as a science-policy interface, other activities and forms of authority matter and shape the order and conduct of an organisation, which can only be identified through detailed study. The actor, activities and forms of authority framework makes it possible to take an organisation apart, to look beyond the forms of authority it may be recognised for and to identify empirically the actors and authorities that matter in shaping its practices and products.

8.3 Government Participation and Power

Studying the IPCC through the actors, activities and forms of authority framework reveals the extent of government involvement in the practice of writing and the symbolic power of some over its conduct. It also illuminates the extent of struggle in the practice of approval and increased level of participation by some developing countries during the final stage of writing, where the assessment's key findings are reformulated for presentation to the world and impact on UNFCCC negotiations. Identifying the extent of member government involvement in the practice of writing required following the assessment along the pathway of its production. It was when documenting the decision to repeat the assessment cycle, the election of a new bureau and the approval of the outline that the activities and avenues for governments to influence the direction and content of the next assessment became apparent.

Chapter 5 reveals the potential for the bureau election to distribute capital and structure the order of relations in the panel, which explains the extent of back channel discussion and manoeuvring documented in WikiLeaks (Section 5.2). Bureau members may sit alongside the national delegation during panel proceedings and the delegate is able to attend bureau meetings, where panel decision-making is discussed and decisions rehearsed. This is important for developed and developing country member governments. For developing country members, it enables them to expand the expertise within their delegation and potentially double their capacity to invest in and undertake IPCC activities, as the travel expenses of both the bureau member and delegate are funded. For all member governments, it enables greater access to and knowledge of the assessment process in practice, as well as the opportunity to build and extend social relations across the bureau, the WG TSUs, the secretariat and other panel members during the smaller, more intimate bureau meetings. As a result, bureau membership enables the accumulation of valuable forms of social and cultural capital, which translate into symbolic power during intergovernmental approval.

It is the approval of the report outline that is most revealing of member government's capacity for structuring the direction and content of the next assessment. The scoping and approval of the report outline serve the dual function of ensuring the next assessment is relevant to its main stakeholders – member governments – and that the co-chairs vision is to some degree aligned with the government's expectations and political concerns in and for the final product. The stakes for member governments in the content of the next report become apparent during the outline's approval, as certain concepts and terms are identified by governments as requiring assessment or there is an attempt to remove them from the outline altogether. This was evident in the struggle over the identification of developed and developing categorisation for assessment in WGIII's outline in the AR5. The Chinese and Saudi Arabian delegations were again careful to ensure this was not inadvertently introduced into the outline for the AR6. However, it is not just terms directly associated with UNFCCC negotiations that can become objects of struggle and requests for removal, as the identification of black carbon in WGI's assessment for the AR5 indicates. Any scientific term or object that shapes global understanding and calculations of the effects of a gas, aerosol or particle on atmospheric warming has the potential to influence climate negotiations and government's policy response in and through the IPCC's practice of writing.

Once the outline is approved, the next activity for governments is the nomination of authors (Section 6.1). Zooming in on the focal point's role and the government's institutionalised process for identifying authors reveals the asymmetry in capacity between developed and developing countries in undertaking IPCC activities. Ninety percent of developed countries surveyed submitted nominations and identified the institutional processes for raising awareness in the national scientific community (IPCC 2009n). This compares to half of developing and EIT country focal points that were surveyed (IPCC 2009n). The first report by the special committee on developing country participation, published in 1992, indicated that the degree of co-ordination between departments and ministries and the 'manpower' [sic] 'to receive, communicate and disseminate information' was not available in most developing countries (IPCC 1992b: 157). The asymmetry in capacity to fulfil the necessary IPCC activities to meaningfully participate and impact the IPCC's practice of writing has continued across assessment cycles, as is apparent from the limited developing country participation in the expert and government review of reports (6.4 and 7.2). Not undertaking a government review has a double impact on developing countries. First, government actors and expertise within these countries are not able to identify the gaps and sources of knowledge necessary for broadening the assessment and ensuring its relevance to their national needs. Second, it is through the review that governments become familiar with and expert on the content of the report and develop a national position on the text. This ensures informed and focused interventions during the approval and concrete proposals that are more likely to be heard and have an impact on the writing of the SPM.

Analysing each activity as an element within the practice of writing makes these patterns of asymmetry and their impacts apparent. Each activity in an assessment's production enables governments to involve and invest in this global attempt to write climate change, ensuring the product is relevant to the national interests and needs for and from climate knowledge. Nominating authors and conducting a government review of the emerging text are not simply avenues to shape the content; they are avenues to know the content and build capacity on and for its re-writing. Through the activity of reviewing, member governments have the potential to learn the latest knowledge on climate change, assess and disseminate its across relevant departments, develop an informed position and to ensure the final product is relevant to national climate policy needs. It is those countries that have the resources to invest, fulfil the government activities and gain knowledge through their undertaking that the IPCC's practice of writing best serves.

Analysing the IPCC makes it apparent that not all member governments are equal in their capacity to influence bureau elections, the outline of the next assessment, its scientific content through the participation of the national scientific community or the wording of the report's key findings. There are vast asymmetries in every stage and element of the IPCC's practice of writing. Bourdieu's concept of capital is critical to revealing the interrelationship and dependency between IPCC participation and economic resources, which are the condition for any country to be present at a meeting, to learn the process and become a meaningful participant within it. However, the resources to attend the meeting, as important as they are in enabling presence, are not sufficient to create a meaningful participant. To understand how symbolically powerful member governments emerge, I have documented the history of the emergence of the IPCC and the cultural foundations of the organisation that those leading the process lay. This is important because it identifies the properties that are valued by an organisation and order relations in that social space, instantly empowering actors embodying this way of being, knowing and doing and designating as inappropriate to the style and conduct of work those that do not.

The cultural foundations of the IPCC elevated scientific and technical modes of knowledge and expertise as the valued properties in members and as ways of organising proceedings. This already provides some explanation for the persistence of the developed and developing country divide within the IPCC and the organisation's ability to meaningfully challenge the asymmetries of participation. Unlike the international scientific actors leading the process, many developing countries did not identify climate change as a scientific and technical issue, but an issue of development. Initial assessment quickly established that developing countries did not have either the scientific/technical expertise on climate change or the resources to attend multiple meetings across the world, through which this expert capacity and knowledge of the emerging process could be developed (IPCC 1992b). From the outset, this had profound effects on some developing countries willingness to accept the IPCC as the basis of knowledge for negotiating climate change (Hughes 2015). And an even greater and lasting impact on all developing countries capacity to become meaningful participants in the IPCC's practice of writing.

8.4 Imprinting Order

What is the imprint of order on the IPCC's practice of writing and most critically, its' products? Carried through the scientific, political and administrative activities of writing climate change and imprinted on its product are the social order of relations of its making. This social order is a product of the distribution of economic, social, scientific and political resources that enable some actors to leave a greater mark on naming climate change than others. This distribution of resources is not unique to the IPCC, although within the IPCC there are uniquely valued properties, it is contiguous to and a reflection of the global distribution of resources.

What we learn from the book is that meaningful participation in the IPCC is resource intensive. On the author and assessment side, interest in the IPCC is dependent on having the national resources to invest in generating and supporting scientific knowledge production and the related institutions and infrastructure

(Chapter 6). A national author's capacity to participate is dependent on time and access to literature. This brings the focus back on national research capacity and the infrastructure and supporting institutions that climate knowledge production is dependent upon (libraries, laboratories, WiFi, computers, computing power, instruments, software, etc.), plus research support for authors to schedule sufficient time for the assessment. On the government side, interest in the panel requires having a designated focal point that preferably remains constant over time, attends all meetings and has the necessary human resources to undertake IPCC activities through which the process is learned, capital accumulated and symbolic power gained to imprint on the practice of writing. These activities include identifying and nominating national experts as authors, having a bureau member, participating in relevant task groups, organising and conducting a government review of the draft report and preparing a well-informed position on the SPM text. This means that IPCC participation is first and foremost dependent on economic capital and a countries capacity to participate is a product of global order.

As the economies and resulting greenhouse gas emissions of some developing countries have increased so has the national scientific capacity, which is observable in the number of authors and co-chairing of an assessment. However, even for countries with growing strength in climate knowledge and expertise, this is not sufficient to significantly imprint on the writing of climate change. It is therefore in the practice of approval that some developing countries emerge as effective writers of climate change. Central to this is the need for consensus (De Pryck 2021), which ensures that a strongly held and spoken objection must be accommodated (with all the clauses identified, about what constitutes the symbolic power to speak and have a strong objection heard). Interestingly, over the last three assessments one of the greatest struggles has emerged over assessment and analysis of developed and developing categories and related responsibilities for emission reductions in the UNFCCC. I have documented this across the approval of the outline (Section 5.4), through review comments (Section 7.2) to the approval of WGIII's contribution to the AR5 (Section 7.3). In the case of the AR5 this resulted in all related content being deleted and careful guarding of the approval of the outline for the AR6 to ensure it remained unassessed.

Another interesting example is the increased participation of Caribbean member governments in the IPCC. In Chapter 7, I document how, combined, Saint Lucia, Saint Kitts and Nevis, and Trinidad and Tobago totalled 6% of the total interventions across the three WGs of the AR6. In the approval sessions, these states frequently intervened to support the authors and to strengthen mentions to issues core to their interests, such as the 1.5 temperature goal, the impacts of climate change, barriers to adaptation, loss and damage and urgency (Bansard, Eni-ibukun and Davenport 2021; Eni-ibukun et al. 2022; Templeton et al. 2022). This reveals just how important the IPCC's practice of approval is for challenging how climate change is named as a collective problem. As developing countries have argued all along, climate change is not just a scientific and technical issue that can be left to scientists and scientific forms of knowledge predominantly produced in the global north to write. Climate change is an issue of development. Core to this problem and knowing this problem, is the global distribution of economic resources and the GHG emissions generated in production and through accumulation. There is order in the IPCC's practice of writing, there is reproduction, and there is also change – as the distribution of resources within the IPCC reflects and facilitates shifts in global order through writing climate change.

8.5 The Implications of the Book

The actors, activities and forms of authority framework of the book makes it possible to study and understand international organisations, like the IPCC, differently. It reveals the social order of relations within an organisation and the values that underpin the distribution of authority and symbolic power, which is a power to shape the conduct of the organisation and its products. The social order of an organisation imprints on its products through the activities that compose it. Mapping the social order and understanding what constitutes that order of relations and with what effect on organisational products makes it possible to identify points to intervene and change.

In the case of the IPCC and as documented in Chapter 4, it becomes apparent that the bureau played a central role in laying the cultural foundations of the IPCC. This included privileging technical and scientific forms of authority in the conduct of the IPCC's work, as well as a concern for broadening the participation of developing countries (Section 4.1). The designation of these values was critical to establishing the IPCC as the leading assessor of climate change and maintaining and strengthening the organisation's symbolic power to name this problem (Sections 3.1 and 4.2). The bureau remains a critical determinant of cultural values in the IPCC. In Chapter 6, I describe how the organisation's focus on gender disparity has significantly increased the number of women in the assessment. I also indicate that while earlier bureau members devalued developing country participants, other bureau members have been key for maintaining the organisational focus on the issue. These actors, with the support of secretariat and TSU actors, have identified measures and mechanisms to create a more inclusive order in the writing of climate change, despite a very limited capacity to shape the broader distribution of social, scientific, political and economic resources. This indicates that within the IPCC, bureau members are well-placed as change makers in the AR7 and beyond. It is in this way that the actors, activities and forms of authority framework makes it possible to both map the order of relations and identify the actors and activities through which the valued properties of the organisation can be challenged and changed.

Understanding the IPCC as practice of writing and the centrality of order in and to the writing of climate change has profound implications for the design of new knowledge processes. This understanding can be used to inform the design of knowledge inputs into treaty-making to prevent and repair global environmental degradation that we as peoples collectively face. The IPCC model has already proven influential in the design of IPBES, which has been referred to as the IPCC of biodiversity (Larigauderie and Mooney 2010). Treaty makers and scholars alike are again looking to the design of the IPCC in negotiations for a new global science-policy body for chemicals and waste (Wang et al. 2021). What are the implications of studying the IPCC as a practice of writing for the design of knowledge bodies for treaty-making purposes? The practice of writing and the actors, activities, forms of authority framework bring to the fore two key dimensions. The first is that science is a site of struggle in agreement-making. The second is that social order matters and emerges from the design of a new organisation, which itself is a product of the existing distributions of economic, social, scientific and political resources or global order.

Despite attempts to separate science and politics and theoretical ideals about the importance of maintaining this separation, science and politics are inseparable in the naming of an environmental issue for collective action. As a result, any assessment body for the purpose of treaty making will be a site of struggle over the meaning of the issue for and in agreement-making. In the establishment of a new body, this struggle plays out over the following dimensions:

1. Where the body will be situated in relation to the treaty – the power it will have in determining decisions and establishing the basis for evaluating implementation.
2. How the knowledge products will be written, on the basis of what rules and procedures, and by what actors, through which activities and on the basis of what authority.

These initial decisions determine the culture of an organisation – the valued properties and their distribution, as recounted in this book for the IPCC. It is this culture that will order relations within the new body and through the assessment activities imprint on the final product. That means, from the outset, the most important facet of a new body is to design for meaningful participation by all members, so that the emerging culture embodies and reflects the multiple worlds and orders that exist in the world and which must all have a place and be preserved in and through collective environmental agreement-making (de la Cadena and Blaser 2018; Vecchione Gonçalves with Hughes 2023). This can be hard to ensure in the political struggle over a new body, in which the powerful seek to privilege the cultural properties of their power. However, if this dominance is left unchecked, there will be nothing left. The order must reflect the diverse ways of knowing, understanding and valuing the Earth so that preserving life on Earth remains central to all activities and valued forms of authority. Perhaps from this, everything else is simple in comparison: practical design and mechanisms to ensure participation by all, as learned through the lessons of the IPCC's practice of writing.

Appendix: Interview Data

	Bureau Members	WMO Region	Location
1	01.07.2010	VI	Neutral
2	26.07.2010	VI	Virtual
3	04.08.2010	VI	Place of work
4	05.08.2010	VI	Virtual
5	07.09.2010	IV	Virtual
6	17.09.2010	I	Virtual
7	09.11.2010	IV	Neutral
8	17.11.2010a	IV	Home
9	26.02.2019a	VI	Place of work
	Focal Point/National Delegate		
10	26.07.2010	VI	Place of work
11	13.12.2010a	IV	Place of work
12	13.12.2010b	V	Virtual
	Technical Support Unit (TSU)		
13	07.07.2010a	VI	Place of work
14	14.07.2010	VI	Virtual
15	25.07.2010	V	Virtual
16	20.01.J2011	IV	Virtual
17	05.10.2010	IV	Place of work
18	05.10.2010	IV	Place of work
19	25.02.2011	IV	Virtual

(*cont.*)

	Bureau Members	WMO Region	Location
20	26.02.2019b	VI	Place of work
21	26.02.2019c	VI	Place of work
	Authors		
22	04.05.2010	VI	Place of work
23	30.06.2010	VI	Place of work
24	30.06.2010	VI	Place of work
25	05.07.2010	VI	Virtual
26	07.07.2010	VI	Neutral
27	14.07.2010	II, V	Virtual
28	27.07.2010	VI	Neutral
29	15.09.2010	IV	Virtual
30	19.09.2010	IV	Virtual
31	21.09.2010	VI	Place of work
32	27.09.2010	IV	Virtual
33	29.09.2010a	IV	Virtual
34	29.09.2010b	VI	Place of work
35	04.10.2010	II, IV	Place of work
36	08.22.2010	IV	Place of work
37	10.11.2010	IV	Place of work
38	20.02.2011	IV	Skype
39	2.02.2011	II	Email questionnaire
40	01.08.2011	V, VI	Neutral

IPCC References

1988

IPCC. (1988). 'Report of the First Session of the WMO/UNEP Intergovernmental Panel on Climate Change (IPCC)'. Geneva, 9–11 November 1988. Available at: www.ipcc.ch/meeting-doc/1st-session-of-the-ipcc-geneva-9-11-november-1988/ (last accessed 15 March 2022).

1989

IPCC. (1989). 'Report of the IPCC Bureau'. Geneva, 6–7 February 1989. Available at: file:///Users/hannahhughes/Downloads/WMOUNEP_IPCC.pdf.pdf (last accessed 27 March 2023).

1990

IPCC. (1990a). 'Report of the Third Session of the WMO/UNEP Panel on Climate Change (IPCC)'. Washington DC, 5–7 February 1990. Available at: www.ipcc.ch/site/assets/uploads/2018/05/first-final-report.pdf (last accessed 23 July 2023).

IPCC. (1990b). 'Policymakers Summary: Prepared by Working Group I'. In Houghton et al. (eds) *Climate Change: The IPCC Scientific Assessment*. Cambridge University Press, Cambridge, United Kingdom.

IPCC. (1990c). 'Report of the Fourth Session of the Intergovernmental Panel on Climate Change (IPCC)'. Sundsvall, Sweden, 27–30 August 1990. Available at: www.ipcc.ch/site/assets/uploads/2018/05/sixth-session-report.pdf (last accessed 23 July 2023).

1991

IPCC. (1991). 'Report of the Fifth Session of the WMO/UNEP Intergovernmental Panel on Climate Change (IPCC)'. Geneva 13–15 March. Available at: www.ipcc.ch/site/assets/uploads/2018/05/fifth-session-report.pdf (last accessed 15 March 2022).

1992

IPCC. (1992a). 'Report of the Eighth Session of the Intergovernmental Panel on Climate Change'. Harare, Zimbabwe, 11–13 November 1992. Available at: www.ipcc.ch/site/assets/uploads/2018/05/eighth-session-report.pdf (last accessed 27 July 2023).

IPCC. (1992b). Policymaker Summary of the IPCC Special Committee on the Participation of Developing Countries. Available at: www.ipcc.ch/site/assets/uploads/2018/05/ipcc_90_92_assessments_far_sc_dev_countries_spm.pdf (last accessed 24 May 2023).

1993

IPCC. (1993). 'Report of the Ninth Session of the Intergovernmental Panel on Climate Change'. Geneva, 29–30 June. Available at: www.ipcc.ch/site/assets/uploads/2018/05/ninth-session-report.pdf (last accessed 27 July 2023).

1995

IPCC. (1995). 'Report of the Eleventh Session of the Intergovernmental Panel on Climate Change'. Rome, Italy, 11–15 December. Available at: www.ipcc.ch/site/assets/uploads/2018/05/eleventh-session-report.pdf (last accessed 23 July 2023).

1997

IPCC. (1997). 'Report of the Thirteenth Session of the Intergovernmental Panel on Climate Change'. Maldives, 22 and 25–28 September. Available at: www.ipcc.ch/meetings/session13/thirteenth-session-report.pdf (last accessed 28 September 2012).

1999

IPCC. (1999). 'Report of the Fifteenth Session of the Intergovernmental Panel on Climate Change'. San Jose, Costa Rica, 15–18 April 1999. Available at: www.ipcc.ch/meetings/session15/fifteenth-session-report.pdf (last accessed 27 July 2023).

2000

IPCC. (2000). Land Use, Land Use Change and Forestry, Cambridge, UK: Cambridge University Press.

2001

IPCC. (2001a). 'Future Work Programme of the IPCC'. IPCC-XVII/Doc. 4. Available at: www.ipcc.ch/site/assets/uploads/2001/04/doc4.pdf (last accessed 27 July 2012).

IPCC. (2001b). 'Future of the Intergovernmental Panel on Climate Change'. IPCC-XVIII/Doc. 4 (a). Available at: www.ipcc.ch/site/assets/uploads/2001/09/doc4a.pdf (last accessed 27 July 2023).

IPCC. (2001c). 'Report of the 17th session of the Intergovernmental Panel on Climate Change'. Nairobi, 4–6 April 2001. Available at: www.ipcc.ch/site/assets/uploads/2018/05/final-report-10.pdf (Last accessed 18 December 2023).

IPCC. (2001d). 'Report of the 18th session of the Intergovernmental Panel on Climate Change'. Wembley, UK, 23–29 September 2001. Available at: www.ipcc.ch/site/assets/uploads/2018/05/final-report-9.pdf (Last accessed 18 December 2023).

2003

IPCC. (2003). 'Report of the 20th Session of the Intergovernmental Panel on Climate Change'. Paris, 19–21 February 2003. Available at: www.ipcc.ch/meetings/session20/final-report.pdf (last accessed 28 September 2012).

2004

IPCC. (2004). 'Progress of the Working Group I towards the IPCC Fourth Assessment Report (AR4)'. IPCC-XXII/Doc. 9. Available at: www.ipcc.ch/meetings/session22/doc9.pdf (last accessed 28 September 2012).

2005

IPCC. (2005). 'Progress Report by Working Group I'. IPCC-XXIV/Doc. 8. Available at: www.ipcc.ch/pdf/session24/doc8.pdf (last accessed 28 September 2012).

2006

IPCC. (2006a). 'Review of IPCC Terms of Reference'. IPCC-XXV/Doc. 8. Available at: www.ipcc.ch/site/assets/uploads/2018/03/doc8-3.pdf (last accessed 13 August 2012).

IPCC. (2006b). 'Draft Rules of Procedures for the Election of the IPCC Bureau and Any Task Force Bureau'. Adopted by the Panel at Its 25th Session, 26–28 April 2006. Available at: www.ipcc.ch/site/assets/uploads/2018/03/doc5-6.pdf (last accessed 27 July 2023).

IPCC. (2006c). 'Progress of Working Group II towards the IPCC Fourth Assessment Report (AR4)'. IPCC-XXV/Doc. 15. Available at: www.ipcc.ch/site/assets/uploads/2018/03/doc15-3.pdf (last accessed 27 July 2023).

IPCC. (2006d). 'IPCC working group I fourth assessment report: expert and government comments on the second order draft – summary for policymakers'. 26 August 2006. Available at: http://pds.lib.harvard.edu/pds/view/7818866 (last accessed 19 December 2023).

2007

IPCC. (2007a). 'Matters Related to the UNFCCC'. Available at: www.ipcc.ch/site/assets/uploads/2018/03/doc7-5.pdf (last accessed 27 July 2023).

IPCC. (2007b). 'Summary for Policymakers'. In Parry et al. (eds) Climate Change 2007: Impacts, Adaptation and Vulnerability. Contribution of Working Group II to the Fourth Assessment Report of the Intergovernmental Panel on Climate Change. Cambridge University Press, Cambridge, UK, 7–22.

IPCC. (2007c). 'Review of IPCC Terms of Reference'. IPCC-XXVI/INF. 2. Available at: www.ipcc.ch/site/assets/uploads/2018/03/doc4-5.pdf (last accessed 27 July 2023).

IPCC. (2007d). 'Report of the 27th Session of the IPCC', Valencia, Spain 12–17 November 2007. Available at: www.ipcc.ch/site/assets/uploads/2018/05/final-report-1.pdf (last accessed 27 July 2023).

2008

IPCC. (2008a). 'Report of the 28th Session of the IPCC, Budapest 9–10 April 2008'. Available at: www.ipcc.ch/site/assets/uploads/2018/05/final-report.pdf (last accessed 3 May 2023).

IPCC. (2008b). 'Future of the IPCC: Compiled Comments from Governments, Authors, Organizations and Bureau Members'. IPCC-XXVIII/INF.1. Available at: www.ipcc.ch/site/assets/uploads/2018/03/inf1-10.pdf (last accessed 27 July 2023).

IPCC. (2008c). 'Future of the IPCC: Synthesis of Comments'. IPCC-XXVIII/Doc. 7. Available at: www.ipcc.ch/site/assets/uploads/2018/03/doc7-7.pdf (last accessed 27 July 2023).

IPCC. (2008d). 'Report of the 29th Session of the IPCC', Geneva, 31 August–4 September 2008. Available at: www.ipcc.ch/site/assets/uploads/2018/05/FINAL_REPT_P_29.pdf (last accessed 3 May 2023).

2009

IPCC. (2009a). 'Report of the 31st Session of the IPCC, Bali, Indonesia, 26–29 October 2009'. Available at: www.ipcc.ch/site/assets/uploads/2018/05/final_report_31.pdf (last accessed 3 May 2023).

IPCC. (2009b). 'Scoping of the IPCC 5th Assessment Report: Background, Cross Cutting Issues and AR5 Synthesis Report'. IPCC-XXXI/Doc. 4. Available at: www.ipcc.ch/site/assets/uploads/2018/03/doc4-10.pdf (last accessed 28 July 2023).

IPCC. (2009c). 'Future IPCC Activities: Reinforcement of the IPCC Secretariat – Report from the Task Group'. IPCC-XXX/Doc.19. Available at: www.ipcc.ch/site/assets/uploads/2018/03/doc19-2.pdf (last accessed 23 July 2023).

IPCC. (2009d). 'Chairman's Vision Paper'. AR5-SCOP/Doc. 2. Available at: https://archive.ipcc.ch/scoping_meeting_ar5/documents/doc02.pdf (last accessed 23 July 2023).

IPCC. (2009e). 'Draft Agenda and Indicative Schedule of Meetings'. AR5-SCOP/Doc. 1. Available at: https://archive.ipcc.ch/scoping_meeting_ar5/documents/doc01.pdf (last accessed 23 July 2023).

IPCC. (2009f). 'Scoping of the IPCC 5th Assessment Report: Summary of Comments on the Draft Scoping Document'. IPCC-XXXI/INF. 5. Available at: www.ipcc.ch/site/assets/uploads/2018/03/inf5-4.pdf (last accessed 28 September 2012).

IPCC. (2009g). 'Report of the 40th Session of the IPCC Bureau'. Geneva, 17–18 September 2009. Available at: www.ipcc.ch/site/assets/uploads/2018/03/bureau40rep.pdf (last accessed 28 September 2012).

IPCC. (2009h). 'Working Group I Contribution to the Fifth Assessment Report (AR5): Background Information'. WG-I:11th /INF.1 (last accessed 28 September 2012).

IPCC. (2009i). 'Working Group II Contribution to the Fifth Assessment Report (AR5): Background Information'. WG-II:9th /INF.1 (last accessed 28 September 2012).

IPCC. (2009j). 'Working Group III Contribution to the Fifth Assessment Report (AR5): Background Information'. WG-III:10th /INF.1 (last accessed 28 September 2012).

IPCC. (2009k). 'Proposed Chapter Outlines for the WG III Contribution to the Fifth Assessment Report (AR5)'. WG-III:10th /Doc.2. Available at: www.ipcc.ch/site/assets/uploads/2018/03/wg3-doc2.pdf (last accessed 28 July 2023).

IPCC. (2009l). 'Agreed Reference Material for the IPCC Fifth Assessment Report' (last accessed 30 August 2012).

IPCC. (2009m). 'Chapter Outline for the WG III Contribution to the Fifth Assessment Report (AR5): Revised Version of WG-III: 10th /Doc.2 Adopted by the Tenth Session of Working Group III'. IPCC-XXXI/Doc. 21. Available at: www.ipcc.ch/site/assets/uploads/2018/03/doc21-1.pdf (last accessed 28 July 2023).

IPCC. (2009n). 'Improving Participation of Developing/EIT Countries in the IPCC: Summary and Recommendations'. IPCC-XXXI/Doc.11. Available at: www.ipcc.ch/site/assets/uploads/2018/03/doc11-8.pdf (last accessed 15 March 2022).

IPCC. (2009o). 'Implementation of Decisions Taken at IPCC 30: Involving Developing/EIT Country Scientists (Decision 7) Questionnaire Results and Analysis on Number of Experts in the Past Assessment Reports'. IPCC-XXXI/INF. 1. Available at: www.ipcc.ch/site/assets/uploads/2018/03/inf1-13.pdf (last accessed 28 July 2023).

2010

IPCC. (2010a). 'Report of the 32nd Session of the IPCC, Busan, Republic of South Korea, 11–14 October 2010'. Available at: www.ipcc.ch/site/assets/uploads/2018/05/final report_32.pdf (last accessed 28 July 2023).

IPCC. (2010b). 'IPCC Trust Fund Programme and Budget'. IPCC-XXXII/Doc.3. Available at: www.ipcc.ch/site/assets/uploads/2018/03/doc03_p32_progr_and_budget.pdf (last accessed 28 July 2023).

IPCC. (2010c). 'Guidance Document for IPCC Government Focal Points on the Nomination of Potential Lead Authors, Coordinating Lead Authors and Review Editors'. Available at: https://archive.ipcc.ch/pdf/ar5/ar5-fp-guidance.pdf (last accessed 28 July 2023).

IPCC. (2010d). '831 Experts Selected for the Fifth Assessment Report'. IPCC Press Release, 23 June 2010. Available at: www.ipcc.ch/site/assets/uploads/2018/04/pr-23june2010.pdf (last accessed 28 July 2023).

IPCC. (2010e). 'The IPCC Fifth Assessment Report: Progress Report of Working Group I'. IPCC-XXXII/Doc. 9. Available at: www.ipcc.ch/site/assets/uploads/2018/03/doc09_p32_wg_I_progress_report.pdf (last accessed 28 July 2023).

IPCC. (2010f). 'The IPCC Fifth Assessment Report: Progress Report of Working Group II'. IPCC-XXXII/Doc. 11. Available at: www.ipcc.ch/site/assets/uploads/2018/03/doc11_p32_wg_II_progress_report.pdf (last accessed 28 July 2023).

IPCC. (2010g). 'The IPCC Fifth Assessment Report: Progress Report of Working Group III'. IPCC-XXXII/Doc. 12. Available at: www.ipcc.ch/site/assets/uploads/2018/03/doc12_p32_wg_III_progress_report.pdf (last accessed 28 July 2023).

IPCC. (2010h). 'Report of the 41st Session of the IPCC Bureau.' Geneva, 19–20 May 2010'. Available at: www.ipcc.ch/site/assets/uploads/2018/03/bureau41rep.pdf (last accessed 28 July 2023).

IPCC. (2010i). 'Letter Inviting Governments to Nominate Experts for the AR5'. Sent to Designated IPCC Focal Points and Ministries of Foreign Affairs (if no focal point has been designated). Geneva, 15 January 2010. Available at: www.ipcc.ch/site/assets/uploads/2018/04/Organisations_Nomination_of_Authors_-for_-the_AR5.pdf (last accessed 17 May 2023).

2011

IPCC. (2011a). 'Report of the 33rd Session of the IPCC, Abu Dhabi, United Arab Emirates, 10–13 May 2011'. Available at: www.ipcc.ch/site/assets/uploads/2018/05/final_report_33.pdf (last accessed 28 July 2023).

IPCC. (2011b). 'Progress Report: Working Group I Contribution to the Fifth Assessment Report'. IPCC-XXXIII/Doc. 16 (last accessed 28 September 2012).

IPCC. (2011c). 'IPCC Fifth Assessment Report (AR5) authors and review editors'. Available at: www.ipcc.ch/site/assets/uploads/2018/03/inf07_p32_ipcc_ar5_authors_review_editors.pdf (last accessed 18 December 2023).

2012

IPCC. (2012a). 'Decisions Takes with Respect to the Review of IPCC Processes and Procedures: Governance and Management'. Available at: www.ipcc.ch/site/assets/uploads/2018/09/IAC_Secretariat_TSU.pdf (last accessed 27 March 2023).

IPCC. (2012b). 'Review of the IPCC Processes and Procedures: Remaining Issues Related to Governance and Management'. Available at: www.ipcc.ch/site/assets/uploads/2018/05/doc23_comments_UNEP_WMO.pdf (last accessed 24 July 2023).

2013

IPCC. (2013). Procedures for the Preparation, Review, Acceptance, Adoption, Approval and Publication of IPCC Reports. Adopted at the Thirty-Seventh Session (Batumi, 14–18 October 2013). Available at: https://archive.ipcc.ch/pdf/ipcc-principles/ipcc-principles-appendix-a-final.pdf (last accessed 9 May 2023).

2014

IPCC. (2014a). IPCC WGIII AR5 Final Government Distribution – Comments on the Final Draft Summary for Policymakers. Available at: https://archive.ipcc.ch/pdf/assessment-report/ar5/wg3/drafts/ipcc_wg3_ar5_fd_comments_summary-for-policymakers.pdf (last accessed 11 July 2023).

IPCC. (2014b). Report of the 40th Session of the IPCC. Copenhagen, Denmark, 27–31 October 2014. Available at: www.ipcc.ch/site/assets/uploads/2018/05/final_report_p40.pdf (last accessed 23 August 2023).

2016

IPCC. (2016a). IPCC Library Facility. Available at: www.ipcc.ch/apps/eventmanager/documents/37/170320160515-Doc.13_LibraryFacility.pdf (last accessed 12 June 2023).

IPCC. (2016b). Participation of Developing Countries in IPCC Activities. Available at: www.ipcc.ch/site/assets/uploads/2018/04/210920161055-INF.4-ParticipationDC.pdf (last accessed 12 June 2023).

2017

IPCC. (2017a). 'Report of the Forty-Sixth Session of the IPCC' Montreal, Canada, 6–10 September, 2017. Available at: www.ipcc.ch/site/assets/uploads/2018/04/final_report_p46.pdf (last accessed 3 May 2023).

IPCC. (2017b). 'Scoping of the IPCC Sixth Assessment Report (AR6): Background, Cross Cutting Issues and the AR6 Synthesis Report'. Available at: www.ipcc.ch/site/assets/uploads/2018/04/040820171122-Doc.-6-SYR_Scoping.pdf (last accessed 24 October 2019).

IPCC. (2017c). 'Proposed Chapter Outlines of the Working Group III Contribution to the IPCC Sixth Assessment Report (AR6)'. Available at: www.ipcc.ch/site/assets/uploads/2018/04/040820170317-WGIII_doc2_chapter_outline.pdf (last accessed 11 May 2023).

2018

IPCC. (2018a). 'Summary for Policymakers'. In: 'Global Warming of 1.5°C. An IPCC Special Report on the Impacts of Global Warming of 1.5°C above Pre-Industrial Levels and Related Global Greenhouse Gas Emission Pathways, in the Context of Strengthening the Global Response to the Threat of Climate Change, Sustainable Development, and Efforts to Eradicate Poverty' [Masson-Delmotte, V., P. Zhai, H.-O. Pörtner, D. Roberts, J. Skea, P.R. Shukla, A. Pirani, W. Moufouma-Okia, C. Péan, R. Pidcock, S. Connors, J.B.R. Matthews, Y. Chen, X. Zhou, M.I. Gomis, E. Lonnoy, T. Maycock, M. Tignor, and T. Waterfield (eds.)].

IPCC. (2018b). 'Report of the Forty-Eighth Session of the IPCC'. Incheon, Republic of Korea, 1–5 October 2018. Available at: www.ipcc.ch/site/assets/uploads/2018/12/final_report_p48.pdf (last accessed 22 May 2023).

IPCC. (2018c). 'Report on the Selection of Coordinating Lead Authors, Lead Authors, and Review Editors of Working Group I Contribution to the Sixth Assessment Report'. Available at: www.ipcc.ch/site/assets/uploads/2018/04/070320180433-INF.-5authors-WGI.pdf (last accessed 22 May 2023).

IPCC. (2018d). Report on the Selection of Coordinating Lead Authors, Lead Authors, and Review Editors of Working Group II Contribution to the Sixth Assessment Report. Available at: www.ipcc.ch/site/assets/uploads/2018/04/070320180433-INF.-2authors-WGII.pdf (last accessed 22 May 2023).

IPCC. (2018e). Report on the Selection of Coordinating Lead Authors, Lead Authors and Review Editors of Working Group III Contribution to the Sixth Assessment Report. Available at: www.ipcc.ch/site/assets/uploads/2018/04/070320180433-INF.-3authors-WGIII.pdf (last accessed 23 May 2023).

IPCC. (2018f). Gender Considerations in the Work of the Intergovernmental Panel on Climate Change. Available at: www.ipcc.ch/site/assets/uploads/2018/04/130220180354-Doc.-7-Gender.pdf (last accessed 25 March 2023).

IPCC. (2018g). Code of Conduct for Meetings and Events of the Intergovernmental Panel on Climate Change. Available at: www.ipcc.ch/site/assets/uploads/2019/07/Code_of_Conduct.pdf (last accessed 2 July 2023).

IPCC. (2018h). Participation of Developing Countries in IPCC Activities. Available at: www.ipcc.ch/site/assets/uploads/2018/04/120220180332-Doc.-4-Part.DVCs_.pdf (last accessed 12 June 2023).

IPCC. (2018i). 'Organisation of the Future Work of the IPCC in Light of the Global Stocktake' Available at: www.ipcc.ch/apps/eventmanager/documents/52/180920180353-INF2Corr.2GlobalStocktake.pdf (last accessed 18 December 2023).

2019

IPCC. (2019). Report from the IPCC Task Force on Gender. Available at: www.ipcc.ch/site/assets/uploads/2019/01/110520190810-Doc.-10-Rev.1TG-Gender.pdf (last accessed 23 May 2023).

2021

IPCC. (2021). IPCC Factsheet: How Does the IPCC Review Process Work? Available at: www.ipcc.ch/site/assets/uploads/2021/07/AR6_FS_review_process.pdf (last accessed 10 June 2023).

2022

IPCC. (2022). Summary for Policymakers. In: *Climate Change 2022: Mitigation of Climate Change*. Contribution of Working Group III to the Sixth Assessment Report of the Intergovernmental Panel on Climate Change [P.R. Shukla, J. Skea, R. Slade, A. Al Khourdajie, R. van Diemen, D. McCollum, M. Pathak, S. Some, P. Vyas, R. Fradera, M. Belkacemi, A. Hasija, G. Lisboa, S. Luz, J. Malley (eds.)]. Cambridge University Press, Cambridge, UK and New York, NY, USA.

2023

IPCC. (2023a). 'Draft Report of the Fifty-Eighth Session of the IPCC'. Interlaken, Switzerland, 13–17 March 2023. Available at: https://apps.ipcc.ch/eventmanager/documents/81/210720230801-Doc.%202,%20Rev.%201-%20Draft%20Report%20IPCC-58.pdf (last accessed 23 July 2023).

IPCC. (2023b). 'Briefing note – IPCC election procedures' Available at: www.ipcc.ch/site/assets/uploads/2023/07/IPCC_election_procedures-final.pdf (last accessed 18 December 2023).

Website

IPCC. (n.d.). IPCC Focal Points. Available at: www.ipcc.ch/apps/contact/interface/focalpoints.php (last accessed 25 September 2019).

IPCC. (n.d.). Selection of Authors for IPCC Sixth Assessment Report. Available at: www.ipcc.ch/2018/04/06/ar6-author-selection/ (last accessed 22 May 2023).

Other References

Abrahamsen, R. & Williams, M. C. (2011). *Security beyond the State: Private Security in International Politics*. Cambridge: Cambridge University Press.

Adler, E. (1992). The Emergence of Cooperation: National Epistemic Communities and the International Evolution of the Idea of Nuclear Arms Control. *International Organization*, 46, 101–45.

Adler, E. & Haas, P. M. (1992). Epistemic Communities, World-Order, and the Creation of a Reflective Research-Program – Conclusion. *International Organization*, 46, 367–90.

Adler, C. E. & Hadorn G. H. (2014). The IPCC and Treatment of Uncertainties: Topics and Sources of Dissensus: IPCC: Topics and Sources of Dissensus. *Wiley Interdisciplinary Reviews: Climate Change* 5(5):663–76.

Adler, E. & Bernstein, S. (2005). Knowledge in Power: The Epistemic Construction of Global Governance. In M. Barnett & R. Duvall, eds., *Power and Global Governance*. Cambridge: Cambridge University Press.

Adler-Nissen, R. ed. (2013). *Bourdieu in International Relations: Rethinking Key Concepts in IR*. New York: Routledge.

Adler-Nissen, R. (2014). Symbolic Power in European Diplomacy: The Struggle between National Foreign Services and the EU's External Action Service. *Review of International Studies*, 40(4), 657–81.

Adler-Nissen, R. & Drieschova, A. (2019). Track-Change Diplomacy: Technology, Affordances, and the Practice of International Negotiations. *International Studies Quarterly*, 63(3), 531–45.

Adler-Nissen, R. & Pouliot, V. (2014). Power in Practice: Negotiating the International Intervention in Libya. *European Journal of International Relations*, 20(4), 889–911.

Agrawala, S. (1998a). Context and Early Origins of the Intergovernmental Panel on Climate Change. *Climatic Change*, 39, 605–20.

Agrawala, S. (1998b). Structural and Process History of the Intergovernmental Panel on Climate Change. *Climatic Change*, 39, 621–42.

Agrawal, S. (1999). Early Science-Policy Interactions in Climate Change: Lessons from the Advisory Group on Greenhouse Gases. *Global Environmental Change and Policy Dimensions*, 9, 157–69.

Allan, B. B. (2017). Producing the Climate: States, Scientists, and the Constitution of Global Governance Objects. *International Organization* 71(1):131–62.

Allan, A., Antonich, B., Bansard, J., Browne, K., Jones, N., & Luomi, M. (2018). Summary of the Katowice Climate Change Conference: 2–15 December 2018. *Earth Negotiations Bulletin*, 12 No. 747. Available at: http://enb.iisd.org/climate/cop24/enb/

Altbach, P. G. (2004). Globalisation and the University: Myths and Realities in an Unequal World. *Tertiary Education and Management*, 10(1), 3–25.

Axelrod, M. (2011). Climate Change and Global Fisheries Management: Linking Issues to Protect Ecosystems or to Save Political Interests? *Global Environmental Politics*, 11(3), 64–84.

Bagley, K. (2012). Climate Scientists Face Organized Harassment in US. *Bloomberg*, 10 September 2012. www.bloomberg.com/news/2012-09-10/climate-scientists-face-organized-harassment-in-u-s-.html (last accessed 10 September 2012).

Bansard, J., Eni-ibukun, T. A., & Davenport, D. (2021). Summary of the 54th Session of the Intergovernmental Panel on Climate Change and the 14th Session of Working Group I: 26 July–6 August 2021. *Earth Negotiations Bulletin*, 12(794), 1–27.

Barnett, A. (2008). IPCC Elections: Close Contests. *Nature Reports Climate Change*, 24 August 2010. Available at: www.nature.com/climate/2008/0810/full/climate.2008.95.html (last accessed 28 April 2010).

Barnett, M. N. & Finnemore, M. (1999). The Politics, Power, and Pathologies of International Organizations. *International Organization*, 53, 699–732.

Bauer, S. (2006). Does Bureaucracy Really Matter? The Authority of Intergovernmental Treaty Secretariats in Global Environmental Politics. *Global Environmental Politics*, 6, 23–49.

van Bavel, B., MacDonald, J. P., & Dorough, D. S. (2022). Indigenous Knowledge Systems. In K. De Pryck & M. Hulme, eds., *A Critical Assessment of the Intergovernmental Panel on Climate Change*. Cambridge: Cambridge University Press, 116–25.

BBC Four. (2021). Climategate: Science of a Scandal. BBC. Available at: www.bbc.co.uk/programmes/m000b8p2 (last accessed 22 July 2023).

Beck, S., Forsyth, T., Kohler, P. M., Lahsen, M., & Mahony, M. (2016). The Making of Global Environmental Science and Politics Silke Beck. In U. Felt, R. Fouché, C. A. Miller, & L. Smith-Doerr, eds., *The Handbook of Science and Technology Studies 4th Edition*. Cambridge, MA: MIT Press, 1059–86.

Beck, S. & Mahony, M. (2018). The IPCC and the New Map of Science and Politics. *Wiley Interdisciplinary Reviews: Climate Change*, 9, n/a–n/a.

van Beek, L., Oomen, J., Hajer, M., Pelzer, P., & van Vuuren, D. (2022). Navigating the Political: An Analysis of Political Calibration of Integrated Assessment Modelling in Light of the 1.5 °C Goal. *Environmental Science & Policy*, 133, 193–202.

Bernstein, S. (2000). Ideas, Social Structure and the Compromise of Liberal Environmentalism. *European Journal of International Relations*, 6, 464–512.

Bernstein, S. (2001). *The Compromise of Liberal Environmentalism*. New York: Columbia University Press.

Bernstein, S., Betsill, M., Hoffmann, M., & Paterson, M. (2010). A Tale of Two Copenhagens: Carbon Markets and Climate Governance. *Millennium: Journal of International Studies*, 39, 161–73.

Bhandari, M. P. (2020). *Getting the Climate Science Facts Right: The Role of the IPCC* (1st ed.). Denmark: River Publishers. www.riverpublishers.com/whoweare.php

Biermann, F. (1999). Big Science, Small Impacts – In the South? The Influence of International Environmental Information Institutions on Policy-Making in India. ENRP Discussion Paper E-99-12, Kennedy School of Government, Harvard University. Available at www.hks.harvard.edu/gea/pubs/e-99-12.htm (last accessed 22 August 2012).

Biermann, F. (2000). Science as Power in International Environmental Negotiations: Global Environmental Assessments between North and South. Belfer Center for Science and International Affairs (BCSIA) Discussion Paper 2000–17, Environment and Natural Resources Program, Kennedy School of Government Harvard University 2000. www.hks.harvard.edu/gea/pubs/2000-17.htm (last accessed 22 August 2012).

Biermann, F. (2002). Institutions for Scientific Advice: Global Environmental Assessments and Their Influence in Developing Countries. *Global Governance*, 8, 195–219.

Biermann, F. & B. Siebenhüner. (2009). *Managers of Global Change: The Influence of International Environmental Bureaucracies*. Cambridge, MA; London: MIT Press.

Biermann, F. & Möller, I. (2019). Rich Man's Solution? Climate Engineering Discourses and the Marginalization of the Global South. *International Environmental Agreements*, 19, 151–67.

Bigo, D. (2006). Global (in)Security: The Field of the Professionals of Unease Management and the Ban-Opticon. In J. Solomon & N. Sakai, eds., *Traces: A Multilingual Series of Cultural Theory*. Hong Kong: Hong Kong University, 109–57.

Bigo, D. (2011). Pierre Bourdieu and International Relations: Power of Practices, Practices of Power. *International Political Sociology*, 5, 225–58.

Bigo, D & Madsen, M. R. (2011). Introduction to Symposium 'A Different Reading of the International': Pierre Bourdieu and International Studies. *International Political Sociology*, 5, 219–24.

Bjurström, A & Polk, M. (2011). Physical and Economic Bias in Climate Change Research: A Scientometric Study of IPCC Third Assessment Report. *Climatic Change*, 108, 1–22.

Black, R. (2009). E-mail Row 'to Affect Copenhagen'. BBC (December 3 2009). Available at: http://news.bbc.co.uk/1/hi/sci/tech/8392611.stm (last accessed 13 July 2023).

Blicharska, M., Smithers, R. J., Kuchler, M., Agrawal, G. K., Gutiérrez, J. M., Hassanali, A., Huq, S., Koller, S. H., Marjit, S., Mshinda, H. M., Masjuki, H. H., Solomons, N. W., Staden, J. V., & Mikusiński, G. (2017). Steps to Overcome the North–South Divide in Research Relevant to Climate Change Policy and Practice. *Nature Climate Change*, 7(1), 21–27.

Bodansky, D. M. (1993). The United Nations Framework Convention on Climate Change: A Commentary. *Yale Journal of International Law*, 18, 451–558.

Boehmer-Christiansen, S. (1994a). Global Climate Protection Policy: The Limits of Scientific Advice Part 1. *Global Environmental Change*, 4(2), 140–59.

Boehmer-Christiansen, S. (1994b). Global Climate Protection Policy: The Limits of Scientific Advice Part 2. *Global Environmental Change*, 4(2), 185–200.

Boehmer-Christiansen, S. (1995a). Britain and the International Panel on Climate Change: The Impacts of Scientific Advice on Global Warming Part I: Integrated Policy Analysis and the Global Dimension. *Environmental Politics*, 4, 1–18.

Boehmer-Christiansen, S. (1995b). Britain and the International Panel on Climate Change: The Impacts of Scientific Advice on Global Warming Part II: The Domestic Story of the British Response to Climate Change. *Environmental Politics*, 4, 175–96.

Boehmer-Christiansen, S. & Skea, J. (1994). The Operation and Impacts of the Intergovernmental Panel on Climate Change: Results of a Survey of Participation and Users, STEEP Discussion Paper, Brighton. SPRU.

Bolin, B. (2007). *A History of the Science and Politics of Climate Change: The Role of the Intergovernmental Panel on Climate Change*. Cambridge: Cambridge University Press.

Booker, C. (2009). The Questions Dr Pachauri Still Has to Answer. *Telegraph*, 26 December 2009. Available at: www.telegraph.co.uk/comment/columnists/christopherbooker/6890839/The-questions-Dr-Pachauri-still-has-to-answer.html (last accessed 4 January 2010).

Borland, R., Morrel, R., & Watson, V. (2018). Southern Agency: Navigating Local and Global Imperatives in Climate Research. *Global Environmental Politics*, 18(3), 47–65.

Borione, D. & J. Ripert. (1994). Exercising Common but Differentiated Responsibility. In M. Mintzer & J. A. Q. Leonard eds., *Negotiating Climate Change: The Inside Story of the Rio Convention*. Cambridge: Cambridge University Press, 77–96.

Bourdieu, P. (1986a). *Distinction: A Social Critique of the Judgement of Taste*. London: Routledge & Kegan Paul, 1984 (1986 [printing]).

Bourdieu, P. (1986b). The Forms of Capital. In J. G. Richardson, ed., *R.J Handbook of Theory and Research for the Sociology of Education*. Westport, CT: Greenwood, 241–58.

Bourdieu, P. (1988). *Homo Academicus*. Cambridge: Polity Press in Association with Basil Blackwell.

Bourdieu, P. (1989). Social Space and Symbolic Power. *Sociological Theory*, 7(1), 14–25.

Bourdieu, P. (1990). *The Logic of Practice*. Cambridge: Polity, 241–58

Bourdieu, P. (2001). [1998] *Masculine Domination*. Translated by R. Nice. Stanford: Stanford University Press.

Bourdieu, P. (2007). *Sketch for a Self-Analysis*. Cambridge: Polity.

Bourdieu, P. (1991). *Language and Symbolic Power*. Edited and Introduced by J. B. Thompson. Cambridge: Polity.

Bourdieu, P. & Wacquant, L. J. D. (1992). *Invitation to Reflexive Sociology*. Chicago & London: University of Chicago Press.

Bowen, K. J., Friel, S., Ebi, K., Butler, C. D., Miller, F., & McMichael. J. D. (2012). Governing for a Healthy Population: Towards an Understanding of How Decision-Making Will Determine Our Global Health in a Changing Climate. *International Journal of Environmental Research and Public Health*, 9, 55–72.

Boyle, S. & Ardill, J. (1989). *The Greenhouse Effect: A Practical Guide to Our Changing Climate*. Kent, England: New English Library.

Brenton, T. (1994). *The Greening of Machiavelli: The Evolution of International Environmental Politics*. London: Royal Institute of International Affairs, Energy and Environmental Programme: Earthscan.

Broome, J. (2014). A Philosopher at the IPCC. *The Philosophers' Magazine Archive*. (2014, July 1).

Broome, J. (2020). Philosophy in the IPCC. In E. Brister & R. Frodeman, eds., *Philosophy for the Real World*. London and New York: Routledge, 95–110.

Brown, D. (2010). Climate Scientist Phil Jones Contemplated Suicide over Data Claims. *Times*, 8 February 2010. Available at: www.timesonline.co.uk/tol/news/environment/article7018484.ece (last accessed 8 February 2010).

Bruce, J. P. (1991). The World Climate Programme's Achievements and Challenges. Proceedings of the Second World Climate Conference. Cambridge University Press, 149–55.

Bruce, J. P., Yi, H., & Haites, E. F. (1996). *Climate Change 1995: Economic and Social Dimensions of Climate Change: Contribution of Working Group III to the Second Assessment Report of the Intergovernmental Panel on Climate Change*. Cambridge: Cambridge University Press.

Butler, S. (2010). Climate Scientists Face Rising Abuse, Threats. 11 June 2011. Available at: www.greenleft.org.au/node/47873 (last accessed 13 June 2011).

Buttel, F. H. (2000). Ecological Modernization as Social Theory. *Geoforum*, 31, 57–65.

de la Cadena, M. with Blaser, M., eds. (2018). *A World of Many Worlds*. Durham: Duke University Press.

Calhoun, C. J., Li-Puma, E., & Postone, M. (1993). *Bourdieu: Critical Perspectives*. Cambridge: Polity.

Campbell, L. M., S. Hagerman, & N. J. Gray. (2014). Producing Targets for Conservation: Science and Politics at the Tenth Conference of the Parties to the Convention on Biological Diversity. Global Environmental Politics 14: 41–63.

Carrington, D. (2010). IPCC Officials Admit Mistake over Melting Himalayan Glaciers. *Guardian* (20 January 2010). Available at: www.guardian.co.uk/environment/2010/jan/20/ipcc-himalayan-glaciers-mistake (last accessed 11 May 2011).

Carter, S., Schulz, A., & Yamineva, Y. (2009). Summary of the 31st Session of the Intergovernmental Panel on Climate Change (IPCC), 26–29 October 2009. *Earth Negotiations Bulletin*, 12, 441.

Chemnick, J. (2018). Oil Kingdom and a 'High Priest' Stall Action for 30 years. E&E News (29 October 2018). Available at: www.eenews.net/articles/oil-kingdom-and-a-high-priest-stall-action-for-30-years/ (last accessed 13 July 2023).

Cohen, S., Demeritt, D., Robinson, J., & Rothman, D. (1998). Climate Change and Sustainable Development: Towards Dialogue. *Global Environmental Change*, 8(4), 341–71.

Cointe, B. (2022). Scenarios. In K. De Pryck & M. Hulme, eds., *A Critical Assessment of the Intergovernmental Panel on Climate Change*. Cambridge: Cambridge University Press, 137–47.

Cointe, B. & Guillemot, H. (2023). 'A History of the 1.5°C Target'. *WIREs Climate Change*. E824.

Collins, P. H. (2015). Intersectionality's Definitional Dilemmas. *Annual Review of Sociology*, 41(1), 1–20.

Collyer, F. M. (2016). Global Patterns in the Publishing of Academic Knowledge: Global North, Global South. *Current Sociology*, 66(1), 56–73.

Compagnon, D. & Bernstein, S. (2017). Nondemarcated Spaces of Knowledge-Informed Policy Making: How Useful Is the Concept of Boundary Organization in IR? *Review of Policy Research*, 34, 812–26.

Confalonieri, U. B. et al. (2007). Human Health. Climate Change 2007: Impacts, Adaptation and Vulnerability. In M. L. Parry, ed., *Contribution of Working Group II to the Fourth Assessment Report of the Intergovernmental Panel on Climate Change*. Cambridge, UK: Cambridge University Press, 391–431.

Conliffe, A. (2011). Combating Ineffectiveness: Climate Change Band Wagoning and the UN Convention to Combat Desertification. *Global Environmental Politics*, 11(3), 44–63.

Connell, R. (2014). Using Southern Theory: Decolonizing Social Thought in Theory, Research and Application. *Planning Theory*, 13(2), 210–23.

Connell, R., Pearse, R., Collyer, F., Maia, J. M., & Morrell, R. (2018a). Negotiating with the North: How Southern-Tier Intellectual Workers Deal with the Global Economy of Knowledge. *The Sociological Review*, 66, 41–57.

Connell, R., Pearse, R., Collyer, F., Maia, J., & Morrell, R. (2018b). Re-making the Global Economy of Knowledge: Do New Fields of Research Change the Structure of North–South Relations? *The British Journal of Sociology*, 69, 738–57.

Van Coppenolle, H., Blondeel, M., & Van de Graaf, T. (2022). Reframing the Climate Debate: The Origins and Diffusion of Net Zero Pledges. *Global Policy*, 14(1), 48–60.

Corbera, E., Calvet-Mir, L., Hughes, H., & Paterson, M. (2016). Patterns of Authorship in the IPCC Working Group III Report. *Nature Climate Change*, 6, 94–99.

de Costello, T. (2009). Smeared Scientists Must Still Mend Their Ways. *Financial Times*, 16 December 2009. Available at: www.ft.com/cms/s/0/0a36733e-ea7a-11de-a9f5-00144feab49a.html?nclick_check=1 (last accessed 30 March 2023).

Craggs, R. & Mahony, M. (2014). The Geographies of the Conference: Knowledge, Performance and Protest. *Geography Compass*, 8(6), 414–30.

Davenport, D., Gutiérrez, M., & Yamineva, M. (2008). Summary of the 28th Session of the Intergovernmental Panel on Climate Change (IPCC), 9–10 April 2008. *Earth Negotiations Bulletin*, 12, 363.

David-Chavez, D. M. & Gavin, M. C. (2018). A Global Assessment of Indigenous Community Engagement in Climate Research. *Environmental Research Letters*, 13, 123005.

Death, C. (2011). Summit Theatre: Exemplary Governmentality and Environmental Diplomacy in Johannesburg and Copenhagen. *Environmental Politics*, 20(1), 1–19.

DECC. (2010). 'List of Applicants to the IPCC 5th Assessment Report Submitted by the UK for Consideration by the IPCC'. Available at: www.decc.gov.uk/assets/decc/what%20we%20do/global%20climate%20change%20and%20energy/tackling%20climate%20change/intl_strategy/ipccreports/1_20100324114204_e_@@_listapplicantsuktoipcc.pdf (last accessed 31 August 2012).

Demeritt, D. (2001). The Construction of Global Warming and the Politics of Science. *Annals of the Association of American Geographers*, 91, 307–37.

Depledge, J. (2006). The Opposite of Learning: Ossification in the Climate Change Regime. *Global Environmental Politics*, 6, 1–22.

Depledge, J. (2007). A Special Relationship: Chairpersons and the Secretariat in the Climate Change Negotiations. *Global Environmental Politics*, 7, 45–68.

Depledge, J. (2008). Striving for No: Saudi Arabia in the Climate Change Regime. *Global Environmental Politics*, 8, 9–35.

De Pryck, K. (2021). Intergovernmental Expert Consensus in the Making: The Case of the Summary for Policy Makers of the IPCC 2014 Synthesis Report. *Global Environmental Politics*, 21(1), 108–29.

De Pryck, K. (2022). Governmental Approval. In K. De Pryck & M. Hulme, eds., *A Critical Assessment of the Intergovernmental Panel on Climate Change*. Cambridge: Cambridge University Press, 187–96.

De Souza, M. (2010). Climate Role Helps National Image: Memo. *Postmedia News*, 17 December 2010. Available at: www.montrealgazette.com/technology/Climate+role+helps+national+image+memo/3990564/story.html (last accessed 19 December 2010).

Dhillon, C. M. (2020). Indigenous Feminisms: Disturbing Colonialism in Environmental Science Partnerships. *Sociology of Race and Ethnicity*, 6(4), 483–500.

Dhillon, C. M. (2022). Indigenous-Settler Climate Change Boundary Organizations Contending with U.S. Colonialism. *American Behavioral Scientist*, 66(7), 937–73.

Drieschova, A. & Bueger, C. (2022). Conceptualising International Practices. In A. Drieschova, C. Bueger, & T. Hopf, eds., *Conceptualizing International Practices: Directions for the Practice Turn in International Relations*. Cambridge: Cambridge University Press, 3–27.

Dubash, N. K., Fleurbaey, M., & Kartha, S. (2014). Political Implications of Data Presentation. *Science*, 345(6192), 36–37.

Dunlap, R. E. & McCright, A. M. (2015). Challenging Climate Change: The Denial Countermovement. In R. E. Dunlap & R. J. Brulle, eds., *Climate Change and Society: Sociological Perspectives*. Oxford: Oxford University Press, 300–32.

Eagleton-Pierce, M. (2013). *Symbolic Power in the World Trade Organization*. Oxford: Oxford University Press.

Edenhofer, O. & Minx, J. (2014). Mapmakers and Navigators, Facts and Values. *Science*, 345(6192), 37–38.

Edwards, P. N. (1999). Global Climate Science, Uncertainty and Politics: Data-Laden Models, Model-filtered Data. *Science as Culture*, 8(4), 437–72.

Edwards, P. N. (2001). Representing the Global Atmosphere: Computer Models, Data, and Knowledge about Climate Change. In C. Miller & P. Edwards, eds., *Changing the Atmosphere: Expert Knowledge and Global Environmental Governance*. Cambridge: MIT Press, 219–46.

Edwards, P. N. (2010). *A Vast Machine: Computer Models, Climate Data, and the Politics of Global Warming*. Cambridge, MA; London: MIT Press.

Edwards, P. N. (2022). Peer Review. In K. De Pryck & M. Hulme, eds., *A Critical Assessment of the Intergovernmental Panel on Climate Change*. Cambridge: Cambridge University Press, 96–104.

Edwards, P. N. & Lahsen, M.(1999). Climate Science and Politics in the US. http://pne.people.si.umich.edu/PDF/PMNPC/USA.pdf (last accessed 23 September 2012).

Edwards, P. N. & Schneider, S. H. (2001). Self-Governance and Peer Review in Science-for-Policy: The Case of the IPCC Second Assessment Report. In C. Miller & P. Edwards eds., *Changing the Atmosphere: Expert Knowledge and Global Environmental Governance*. Cambridge: MIT Press, 219–46.

Eilerpin, J. (2007). U.S., China Got Climate Warnings Toned Down. Washington Post 7 April 2007. www.washingtonpost.com/wp-dyn/content.article/2007/04/06/AR2007040600291.html. (last accessed 16 September 2012).

den Elzen, M. G. J., Höhne, N., Brouns, B., Winkler, H., & Ott, H. E. (2007). Differentiation of Countries' Future Commitments in a Post-2012 Climate Regime: An Assessment of the 'South–North Dialogue' Proposal. *Environmental Science & Policy* 10(3):185–203.

Eni-ibukun, T. A., Davenport, D., Gutiérrez, M., & Mead, L. (2022). Summary of the 55th Session of the Intergovernmental Panel on Climate Change and the 12th Session of Working Group II: 14–27 February 2022. *Earth Negotiations Bulletin*, 12(794), 1–24.

Epstein, C. (2008). *The Power of Words in International Relations: Birth of an Anti-Whaling Discourse*. Cambridge, MA; London: MIT.

Fidler, D. (2010). The Challenges of Global Health Governance. Working Paper for Council of Foreign Relations. Available at: http://ec.europa.eu/health/eu_world/docs/ev_20111111_rd01_en.pdf (last accessed 30 April 2012).

Field, C. B. et al. (2012). *Managing the Risks of Extreme Events and Disasters to Advance Climate Change Adaptation*. A Special Report of Working Groups I and II of the Intergovernmental Panel on Climate Change. Cambridge University Press, Cambridge, UK, and New York, NY, USA, pp. 3–21.

Fisher, D. R & W. R. Freudenburg. (2001). Ecological Modernization and Its Critics: Assessing the Past and Looking toward the Future. *Society & Natural Resources* 14:701–709.

Fligstein, Neil & Doug McAdam. (2011). Toward a General Theory of Strategic Action Fields. *Sociological Theory* 29(1):1–26.

Fligstein, Neil & Doug McAdam. (2014). *A Theory of Fields*. Thousand Oaks, CA: Sage.

Fogel, C. (2005). Biotic Carbon Sequestration and the Kyoto Protocol: The Construction of Global Knowledge by the Intergovernmental Panel on Climate Change. *International Environmental Agreements*, 5, 191–210.

Ford, J. D., Vanderbilt, W., & Berrang-Ford, L. (2012). Authorship in IPCC AR5 and Its Implications for Content: Climate Change and Indigenous Populations in WGII. *Climatic Change*, 113(2), 201–13.

Ford, J. D., Cameron, L., Rubis, J., Maillet, M., Nakashima, D., Willox, A. C., & Pearce, T. (2016). Including Indigenous Knowledge and Experience in IPCC Assessment Reports. *Nature Climate Change*, 6(4), 349–53.

Forster, P. et al. (2007). Changes in Atmospheric Constituents and in Radiative Forcing. In S. Solomon et al. (eds) *Climate Change 2007: The Physical Science Basis. Contribution of Working Group I to the Fourth Assessment Report of the Intergovernmental Panel on Climate Change*. Cambridge, United Kingdom and New York, NY, USA: Cambridge University Press.

Franz, W. E. (1997). The Development of an International Agenda for Climate Change: Connecting Science to Policy. Discussion Paper E-97-07, Kennedy School of Government, Harvard University, August 1997.

Gay-Antaki, M. (2021). Stories from the IPCC: An Essay on Climate Science in Fourteen Questions. *Global Environmental Change*, 71, 102384.

Gay-Antaki, M. & Liverman, D. (2018). Climate for Women in Climate Science: Women Scientists and the Intergovernmental Panel on Climate Change. *Proceedings of the National Academy of Sciences*: 201710271.

Gettelman, A. (2003). The Information Divide in the Climate Sciences. Report of a 2002 Survey, Environmental and Societal Impacts Group, National Center for Atmospheric Research, Boulder, CO, 52 pp. Available at: www.esig.ucar.edu/infodivide (last accessed 22 September 2012).

Gibbs, W. W. (1995). Lost Science in the 3rd-World. *Scientific American*, 273, 92–99.

Gieryn, T. F. (1983). Boundary-Work and the Demarcation of Science from Non-Science: Strains and Interests in Professional Ideologies of Scientists. *American Sociological Review*, 48, 781–95.

Gov.uk (n.d). IPCC Call for UK Experts to Produce the IPCC Sixth Assessment Report. Available at: www.gov.uk/government/publications/ipcc-call-for-uk-experts-to-produce-the-ipcc-sixth-assessment (last accessed 22 May 2023).

Guardian. (2010a). US Embassy Cables: US Lobbied Rajendra Pachauri to Help them Block Appointment of Iranian Scientist. *The Guardian* (6 December 2010). Available at: www.guardian.co.uk/world/us-embassy-cables-documents/168194 (last accessed 8 December 2010).

Guardian. (2010b). US Embassy Cables: Norway Supports US Plan to Block Election of Iranian Climate Scientist. *The Guardian* (6 December 2010). Available at: www .guardian.co.uk/world/us-embassy-cables-documents/166258 (last accessed 8 December 2010).

Guardian. (2010c). US Embassy Cables: Brazil Considers US Plan to Block Election of Iranian Climate Scientist. *The Guardian* (6 December 2010). Available at: www.guardian.co.uk/world/us-embassy-cables-documents/166298 (last accessed 8 December 2010).

Guardian. (2010d). US Embassy Cables: Saudi Arabia Fears Missed Trick on Copenhagen Climate Accord. *The Guardian* (3 December 2010). Available at: www.theguardian .com/world/us-embassy-cables-documents/248643

Guillemot, H. (2022). Climate Models. In K. De Pryck & M. Hulme eds., *A Critical Assessment of the Intergovernmental Panel on Climate Change*. Cambridge: Cambridge University Press, 126–36.

Gupta, J. (2013). *The Climate Change Convention and Developing Countries: From Conflict to Consensus? Dordrecht*, Netherlands: Kluwer.

Gupta, S., Tirpak, D. A., Burger, N., et al. (2007). Policies, Instruments and Co-operative Arrangements. In B. Metz, O. R. Davidson, P. R. Bosch, R. Dave, & L. A. Meyer eds., *Climate Change 2007: Mitigation. Contribution of Working Group III to the Fourth Assessment Report of the Intergovernmental Panel on Climate Change*. Cambridge, UK, and New York: Cambridge University Press.

Gustafsson, K. (2019). Learning from the Experiences of the Intergovernmental Panel on Climate Change: Balancing Science and Policy to Enable Trustworthy Knowledge. *Sustainability*, 11(22), 6533.

Guston, D. H. (1999). Stabilizing the Boundary between US Politics and Science: The Role of the Office of Technology Transfer as a Boundary Organization. *Social Studies of Science*, 29, 87–111.

Guston, D. H. (2001). Boundary Organizations in Environmental Policy and Science: An Introduction. *Science Technology & Human Values*, 26, 399–408.

Gutiérrez, M., Muñoz M., & Johnson, S. S. (2007a). Tenth Session of Working Group I of the Intergovernmental Panel on Climate Change: 29 January–1 February 2007. *Earth Negotiations Bulletin*, 12(319), 1–18.

Gutiérrez, M., Kulovesi, K., & Muñoz, M. (2007b). Eighth Session of Working Group II of the Intergovernmental Panel on Climate Change: 2–6 April 2007. *Earth Negotiations Bulletin*, 12 (320), 1–12.

Gutiérrez, M. et al. (2007c). Ninth Session of Working Group III of the Intergovernmental Panel on Climate Change: 30 April–4 May 2007. *Earth Negotiations Bulletin*, 12 (321), 1–16.

Gutiérrez, M., van Alstine, J., & Yamineva, M. (2011). Summary of the 34th Session of the Intergovernmental Panel on Climate Change: 18–19 December 2012. *Earth Negotiations Bulletin*, 15 (522), 1–12.

Gutiérrez, M., Kosolapova, A., Luomi, M., & Mead, L. (2014). Summary of the 12th Session of Working Group III of the Intergovernmental Panel on Climate Change: 7–12 April 2014. *Earth Negotiations Bulletin*, 12(597), 1–24.

Haas, P. M. (1989). Do Regimes Matter – Epistemic Communities and Mediterranean Pollution Control. *International Organization*, 43, 377–403.

Haas, P. M. (1990). Obtaining International Environmental Protection through Epistemic Consensus. *Millennium: Journal of International Studies*, 19, 347–63.

Haas, P. M. (1992a). Epistemic Communities and International-Policy Coordination – Introduction. *International Organization*, 46, 1–35.

Haas, P. M. (1992b). Climate Change Negotiations. *Environment*, 34, 1–2.

Haas, P. M. (2000). International Institutions and Social Learning in the Management of Global Environmental Risks. *Policy Studies Journal*, 28, 558–75.

Haas, P. M. (2004). When Does Power Listen to Truth? A Constructivist Approach to the Policy Process. *Journal of European Public Policy*, 11, 569–92.

Haas, P. M. (2005). Science and International Environmental Governance. In P. Dauverge ed., *Handbook of Global Environmental Politics*. UK: Edward Elgar Publishing Limited, 383–401.

Haas, P. M. (2008). Climate Change Governance after Bali. *Global Environmental Politics*, 8, 1–7.

Haas, P. M. & McCabe, D. (2001). Amplifiers or Dampeners: International Institutions and Social Learning in the Management of Global Environmental Risks. In, volume edited by the Social Learning Group. *Learning to Manage Global Environmental Risks, volume 1*. Cambridge, MA: MIT Press, 323–48.

Haas, P. M. & Stevens, C. (2011). Organized Science, Usable Knowledge, and Multilateral Environmental Governance. In R. Lidskog & G. Sundkvist (eds) *Governing the Air: The Dynamics of Science, Policy and Citizen Interaction*. Cambridge MA: The MIT Press, 125–164.

Haines, A. (2008). Climate Change and Health Strengthening the Evidence Base for Policy. *American Journal of Preventive Medicine*, 35, 411–13.

Haines, A. et al. (2007). Energy and Health 6 – Policies for Accelerating Access to Clean Energy, Improving Health, Advancing Development, and Mitigating Climate Change. *Lancet*, 370, 1264–81.

Haines, A. et al. (2009). Health and Climate Change 6 Public Health Benefits of Strategies to Reduce Greenhouse-Gas Emissions: Overview and Implications for Policy Makers. *Lancet*, 374, 2104–14.

Hajer, M. A. (1995). *The Politics of Environmental Discourse: Ecological Modernization and the Policy Process*. Oxford: Clarendon Press.

Hall, N. (2015). Money or Mandate?: Why International Organizations Engage with the Climate Change Regime. *Global Environmental Politics*, 15(2), 79–97.

Harris, J. (2009). 'Copenhagen Climate Conference: The Key Players'. *Guardian*, 30 November 2009. Available at: www.guardian.co.uk/environment/2009/nov/30/copenhagen-key-players (last accessed 17 August 2011).

Hart, D. M. & Victor, D. G. (1993). Scientific Elites and the Making of US Policy for Climate-Change Research, 1957–74. *Social Studies of Science*, 23(4), 643–80.

Hashimoto, S. et al. (1990). Human Settlement; The Energy, Transport, and Industrial Sectors; Human Health; Air Quality; and Changes in Ultraviolet-B Radiation. In W. J. Tegart, G. W. Sheldon, & D. C. Griffiths, eds., *Climate Change: The IPCC Impacts Assessment*. Australia: Department of the Arts, Sport, the Environment, Tourism and Territories by the Australian Government Publishing Service, 163–207.

Hecht, A. D. & Tirpak, D. (1995). Framework Agreement on Climate Change – A Scientific and Policy History. *Climatic Change*, 29, 371–402.

Hermansen, E. A. T., Lahn, B., Sundqvist, G., & Øye, E. (2021). Post-Paris Policy Relevance: Lessons from the IPCC SR15 Process. *Climatic Change*, 169(1), 7.

Hilgartner, S. (2000). *Science on Stage: Expert Advice as Public Drama*. Stanford: Stanford University Press.

Hiramatsu, A., Mimura, N., & Sumi, A. (2008). A Mapping of Global Warming Research Based on IPCC AR4. *Sustainability Science*, 3(2), 201–13.

Ho-Lem, C., Zerriffi, H., & Kandlikar, M. (2011). Who Participates in the Intergovernmental Panel on Climate Change and Why: A Quantitative Assessment of the National Representation of Authors in the Intergovernmental Panel on Climate Change. *Global Environmental Change*, 21, 1308–17.

Hoppe, R., Wesselink, A., & Cairns, R. (2013). Lost in the Problem: The Role of Boundary Organisations in the Governance of Climate Change. *Wiley Interdisciplinary Reviews: Climate Change*, 4, 283–300.

Houghton, J. (2002). An Overview of the Intergovernmental Panel on Climate Change (IPCC) and Its Process of Science Assessment. *Issues in Environmental Science and Technology*, 17, 1–20.

Houghton, J. (2008). Madrid 1995: Diagnosing Climate Change. *Nature*, 455, 737–38.

Houghton, J. T., Jenkins, G. J., & Ephraums, J. J. (1990). Climate Change: The IPCC Scientific Assessment. *Published for the Intergovernmental Panel on Climate Change. Cambridge University Press*, 1990 (1991).

Houghton, J. T. et al. (1996). *Climate Change 1995: The Science of Climate Change: Contribution of WG 1 to the Second Assessment Report of the Intergovernmental Panel on Climate Change*. Cambridge: Cambridge University Press.

Houghton, J. T. (2001). *Climate Change 2001: The Scientific Basis: Contribution of Working Group I to the Third Assessment Report of the Intergovernmental Panel on Climate Change*. Cambridge: Cambridge University Press.

Hughes, H. (2015). Bourdieu and the IPCC's Symbolic Power. *Global Environmental Politics*, 15, 85–104.

Hughes, H. (2022). Governments. In K. De Pryck & M. Hulme, eds., *A Critical Assessment of the Intergovernmental Panel on Climate Change*. Cambridge: Cambridge University Press, 79–87.

Hughes, H. (2023). Actors, Activities, and Forms of Authority in the IPCC. *Review of International Studies*, 1–21.

Hughes, H. & Paterson, R. M. (2017). Narrowing the Climate Field: The Symbolic Power of Authors in the IPCC's Assessment of Mitigation. *Review of Policy Research*, 34, 744–66.

Hughes, H. & Vadrot, A. B. M. (2019). Weighting the World: IPBES and the Struggle over Biocultural Diversity. *Global Environmental Politics*, 19, 14–37.

Hughes, H., Vadrot, A., Allan, J. I., Bach, T., Bansard, J. S., Chasek, P., Gray, N., Langlet, A., Leiter, T., Marion Suiseeya, K. R., Martin, B., Paterson, M., Ruiz-Rodríguez, S. C., Wysocki, I. T., Tolis, V., Thew, H., Gonçalves, M. V., & Yamineva, Y. (2021). Global Environmental Agreement-Making: Upping the Methodological and Ethical Stakes of Studying Negotiations. *Earth System Governance*, 10, 100121.

Hughes, H. & Vadrot, A. B. M. (2023). Introduction: A Broadened Understanding of Global Environmental Negotiations. In H. Hughes & A. B. M. Vadrot, eds. (2023). *Conducting Research on Global Environmental Agreement-Making*. Cambridge: Cambridge University Press, 1–22.

Hull, M. S. (2003). The File: Agency, Authority, and Autography in an Islamabad Bureaucracy. *Language & Communication*, 23(3), 287–314.

Hull, M. S. (2012). Government of Paper: The Materiality of Bureaucracy in Urban Pakistan.

Hulme, M. (2008). Geographical Work at the Boundaries of Climate Change. *Transactions of the Institute of British Geographers*, 33(1), 5–11.

Hulme, M. & Mahony, M. (2010). Climate Change: What Do We Know about the IPCC? *Progress in Physical Geography*, 34, 705–18.

IAC. (2010a) Review of IPCC processes and procedures. Report by the InterAcademy Council. IPCC-XXXII/Doc.7. Available at: www.ipcc.ch/site/assets/uploads/2018/03/doc07_p32_report_IAC.pdf (last accessed 03 August 2023).

IAC. (2010b). Responses to the IAC questionnaire. Available at: https://hro001.files.wordpress.com/2010/12/iacreviewcommentsbookmarkedannotated.pdf (last accessed 3 August 2023).

Ibarra, C., Jiménez, G., O'Ryan, R., Blanco, G., Cordero, L., Insunza, X., Moraga, P., Rojas, M., & Sapiains, R. (2022). Scientists and Climate Governance: A View from the South. *Environmental Science & Policy*, 137, 396–405.

Jackson, P. (2009). Pierre Bourdieu. In J. Edkins & N. Vaughan-Williams, eds., *Critical Theorists and International Relations*. London: Routledge, 102–13.

Jaeger, J. & O'Riordan, T. (1996). The History of Climate Change Science and Politics. In J. Jaeger & T. O'Riordan, eds., *Politics of Climate Change: A European Perspective*. London: Routledge, 1–31.

Jasanoff, S. (2004a). The Idiom of Co-production. In S. Jasonoff, ed., *States of Knowledge: The Co-production of Science and Social Order*. International Library of Sociology. London: Routledge, 1–12.

Jasanoff, S. (2004b). Ordering Knowledge, Ordering Society. In S. Jasonoff ed., *States of Knowledge: The Co-production of Science and Social Order*. International Library of Sociology. London: Routledge.

Jinnah, S. (2010). Overlap Management in the World Trade Organization: Secretariat Influence on Trade-Environment Politics. *Global Environmental Politics*, 10, 54–79.

Jinnah, S. (2011a). Climate Change Bandwagoning: The Impacts of Strategic Linkages on Regime Design, Maintenance, and Death Introduction. *Global Environmental Politics*, 11(3), 1–9.

Jinnah, S. (2011b). Marketing Linkages: Secretariat Governance of the Climate-Biodiversity Interface. *Global Environmental Politics*, 11(3), 23–43.

Jinnah, S. (2014). *Post-treaty Politics: Secretariat Influence in Global Environmental Governance*. Cambridge MA: The MIT Press.

Jöns, H. & Hoyler, M. (2013). Global Geographies of Higher Education: The Perspective of World University Rankings. *Geoforum*, 46, 45–59.

Jowitt, J. (2011). The Leaked Climate Science Emails – and What They Mean. *Guardian*, 24 November 2011. Available at: www.guardian.co.uk/environment/2011/nov/24/leaked-climate-science-emails (last accessed 5 September 2012).

Kandlikar, M. & Sagar, A. (1999). Climate Change Research and Analysis in India: An Integrated Assessment of a South-North Divide. *Global Environmental Change-Human and Policy Dimensions*, 9(2), 119–38.

Karlsson, S., Srebotnjak, T., & Gonzales, P. (2007). Understanding the North-South Knowledge Divide and Its Implications for Policy: A Quantitative Analysis of the

Generation of Scientific Knowledge in the Environmental Sciences. *Environmental Science & Policy*, 10(7–8), 668–84.

Kellogg, W. W. (1987). Mankinds Impact on Climate – the Evolution of an Awareness. *Climatic Change*, 10(2), 113–36.

Keohane, R. O. & Victor, D. G. (2011). The Regime Complex for Climate Change. *Perspectives on Politics*, 9, 7–23.

Ketcham, C. (2022). How Scientists from the 'Global South' Are Sidelined at the IPCC. *The Intercept*, posted 17 November 2022. https://theintercept.com/2022/11/17/climate-un-ipcc-inequality/ (last accessed 5 June 2023).

Klenk, N., Fiume, A., Meehan, K., & Gibbes, C. (2017). Local Knowledge in Climate Adaptation Research: Moving Knowledge Frameworks from Extraction to Co-production. *WIREs Climate Change*, 8, e475.

Kouw, M. & Petersen, A. (2018). Diplomacy in Action: Latourian Politics and the Intergovernmental Panel on Climate Change. *Science & Technology Studies*, 31(1), 52–68. Article 1.

Lahn, B. (2021). Changing Climate Change: The Carbon Budget and the Modifying Work of the IPCC. *Social Studies of Science*, 51(1), 3–27.

Lahn, B. (2022). Boundary Objects. In K. De Pryck & M. Hulme, eds., *A Critical Assessment of the Intergovernmental Panel on Climate Change*. Cambridge: Cambridge University Press, 225–33.

Lahn, B. & Sundqvist, G. (2017). Science as a "Fixed Point"? Quantification and Boundary Objects in International Climate Politics. *Environmental Science & Policy*, 67, 8–15.

Lahsen, M. (1998). The Detection and Attribution of Conspiracies: The Controversy over Chapter 8. In G. E. Marcus ed., *Paranoia within Reason: A Casebook on Conspiracy as Explanation. Late Editions 6, Cultural Studies for the End of the Century*. Chicago, IL: University of Chicago Press, Chicago.

Lahsen, M. (2004). Transnational Locals: Brazilian Experiences of the Climate Regime. In S. Jasanoff & M. L. Martello, eds., *Earthly Politics: Local and Global in Environmental Governance*. Cambridge, MA: MIT Press, 159.

Lahsen, M. (2008). Experiences of Modernity in the Greenhouse: A Cultural Analysis of a Physicist 'Trio' Supporting the Backlash against Global Warming. *Global Environmental Change-Human and Policy Dimensions*, 18(1), 204–19.

Langhelle, O. (2000). Why ecological modernisation and sustainable development should not be conflated. *Journal of Environmental Policy and Planning* 2: 303–22.

Latulippe, N. & Klenk, N. (2020). Making Room and Moving Over: Knowledge Co-production, Indigenous Knowledge Sovereignty and the Politics of Global Environmental Change Decision-making. *Current Opinion in Environmental Sustainability. Advancing the Science of Actionable Knowledge for Sustainability*, 42, 7–14.

Larigauderie, A. & Mooney, H. A. (2010). The Intergovernmental Science-Policy Platform on Biodiversity and Ecosystem Services: Moving a Step Closer to an IPCC-Like Mechanism for Biodiversity. *Current Opinion in Environmental Sustainability*, 2(1), 9–14.

Lawler, A. (2002). Pachauri Defeats Watson in New Chapter for Global Panel. *Science*, 296 (5568), 632.

Leander, A. (2011). The Promises, Problems, and Potentials of a Bourdieu-Inspired Staging of International Relations. *International Political Sociology*, 5, 294–313.

Leggett, J. K. (1999). *The Carbon War: Dispatches from the End of the Oil Century*. London: Allen Lane.

Levin, K., Cashore, B., Bernstein, S., & Auld, G. (2009). Playing It Forward: Path Dependency, Progressive Incrementalism, and the 'Super Wicked' Problem of Global Climate Change. *IOP Conference Series: Earth and Environmental Science*, 6(50), 502002.

Levin, K., Cashore, B., Bernstein, S., & Auld, G. (2012). Overcoming the Tragedy of Super Wicked Problems: Constraining Our Future Selves to Ameliorate Global Climate Change. *Policy Sciences*, 45(2), 123–52.

Lidskog, R. & Sundqvist, G. (2015). When Does Science Matter? International Relations Meets Science and Technology Studies. *Global Environmental Politics*, 15(1), 1–20.

Litfin, K. (1994). *Ozone Discourse: Science and Politics in Global Environmental Cooperation*. New York; Chichester. Columbia University Press.

Liverman, D., vonHedemann, N., Nying'uro, P., Rummukainen, M., Stendahl, K., Gay-Antaki, M., Craig, M., Aguilar, L., Bynoe, P., Call, F., Connors, S., David, L., Ferrone, A., Hayward, B., Jayawardena, S., Mai Touray, L., Parikh, J., Pathak, M., Perez, R., Pirani, A., Prakash, A., Textor, C., Tibig, L., Tignor, M., Tuğaç, Ç., Vera, C., & Wagle, R. (2022). Survey of Gender Bias in the IPCC. *Nature*, 602(7895), 30–32.

Livingston, G., Waring, B., Pacheco, L. F., Buchori, D., Jiang, Y., Gilbert, L., & Jha, S. (2016). Perspectives on the Global Disparity in Ecological Science. *BioScience*, 66, 147–55.

Livingston, J. E., Lövbrand, E., & Olsson, J. A. (2018). From Climates Multiple to Climate Singular: Maintaining Policy-Relevance in the IPCC Synthesis Report. *Environmental Science & Policy*, 90, 83–90.

Livingston, J. E. & Rummukainen, M. (2020). Taking Science by Surprise: The Knowledge Politics of the IPCC Special Report on 1.5 Degrees. *Environmental Science and Policy*, 112, 10–16.

Lorenzoni, I. & Harold, J. (2022). Visuals. In K. De Pryck, & M. Hulme, eds., *A Critical Assessment of the Intergovernmental Panel on Climate Change*. Cambridge: Cambridge University Press. 234–43.

Lövbrand, E. (2007). Pure Science or Policy Involvement? Ambiguous Boundary-Work for Swedish Carbon Cycle Science. *Environmental Science and Policy*, 10(1), 39–47.

Lunde, L. (1991). *Science or Politics in the Global Greenhouse? The Development towards Scientific Consensus on Climate Change*. Lysaker: Fridtjof Nansen Institute.

Mach, K. J., Freeman, P. T., Mastrandrea, M. D., & Field, C. B. (2016). A Multistage Crucible of Revision and Approval Shapes IPCC Policymaker Summaries. *Science Advances*, 2(8), e1600421.

Mahony, M. (2014). The Predictive State: Science, Territory and the Future of the Indian Climate. *Social Studies of Science*, 44(1), 109–33.

Mahony, M. (2015). Climate Change and the Geographies of Objectivity: The Case of the IPCC's Burning Embers Diagram. *Transactions of the Institute of British Geographers*, 40, 153–67.

Mason, J. (2007). 'Scientists, Governments Clash over Warming Report' *Reuters* (6 April 2007) at: www.reuters.com/article/2007/04/06/us-globalwarming-scientists-idUSL0649942120070406 (last accessed 29 September 2011)

Masood, E. (1996). Climate Report 'Subject to Scientific Cleansing'. *Nature*, 381 (13 June 1996), 546.

McCright, A. M. & Dunlap, R. E. (2003). Defeating Kyoto: The Conservative Movement's Impact on U.S. Climate Change Policy. *Social Problems*, 50(3), 348–73.

McCright, A. M. & Dunlap, R. E. (2010). Anti-Reflexivity the American Conservative Movement's Success in Undermining Climate Science and Policy. *Theory Culture & Society*, 27, 100–33.

McGrath, M. (2018). 'Climate Change: COP24 Fails to Adopt Key Scientific Report', *BBC News*, 8 December, at www.bbc.co.uk/news/science-environment-46496967 (last accessed 14 August 2019).

McIntyre, S. (2009). Climate Gatekeeping: Michaels and McKitrick 2004. *Climate Audit*, 17 December 2009. Available at: http://climateaudit.org/2009/12/17/climategatekeeping-2/ (last accessed 3 September 2012).

McLaren, C. & Carter, A. (2010). 'UKCDS Workstream Review: Review of Funding and Outputs of UK Research on Climate Change and Development'.

McMichael, A. J. et al. (1996). Human Population Health. In R. Watson et al. eds., *Climate Change 1995: Impacts, Adaptations and Mitigation of Climate Change: Scientific-Technical Analyses*. Cambridge: Cambridge University Press, 561–85.

McMichael, A. J. et al. (2001). Human Health. In J. J. McCarthy, et al. eds., *Climate Change 2001: Impacts, Adaptation and Vulnerability*. Cambridge: Cambridge University Press, 451–86.

Mead, L., Cardenes, I., Gutiérrez, M., & Woods, B. (2017). Summary of the 47th Session of the Intergovernmental Panel on Climate Change: 6–10 September 2017. *Earth Negotiations Bulletin*, 12 (702), 1–15.

Meadows, D. H., Randers, J., & Meadows, D. L. (1972). *The Limits to Growth: A Report for the Club of Rome's Project on the Predicament of Mankind*. New York: New American Library.

Miller, C. (2001a). Scientific Internationalism in American Foreign Policy: The Case of Meteorology, 1947–1958. In C. Miller & P. N. Edwards, eds., *Changing the Atmosphere: Expert Knowledge and Environmental Governance*. Cambridge, MA; London: MIT, 167–208.

Miller, C. (2001b). Hybrid Management: Boundary Organizations, Science Policy, and Environmental Governance in the Climate Regime. *Science Technology and Human Values*, 26(4), 478–500.

Miller, C. (2004). Climate Science and the Making of a Global Political Order. In S. Jasanoff ed., *States of Knowledge: The Co-production of Science and Social Order*. International Library of Sociology. London: Routledge, 46–66.

Mitchell, R. B., W. C. Clark, D. W. Cash, & N. M. Dickson (eds)(2006). *Global Environmental Assessments: Information and Influence*. Cambridge, MA; London: MIT.

Mol, A. P. J. & G. Spaargaren (2000). Ecological Modernization Theory in Debate: A Review. *Environmental Politics* 9:17–49.

Monbiot, G. (2010). 'Climate change email scandal shames the university and requires resignations'. *Guardian*, 2 February 2010. Available at: www.guardian.co.uk/environment/georgemonbiot/2010/feb/02/climate-change-hacked-emails (last accessed 18 December 2023).

Moss, R. H. (2000). Ready for IPCC-2001: Innovation and Change in Plans for the IPCC Third Assessment Report – An Editorial Comment. *Climatic Change*, 45, 459–68.

Nature. (1996). Climate Debate Must Not Overheat. *Nature*, 381, 6583.

Neumann, I. B. (2002). Returning Practice to the Linguistic Turn: The Case of Diplomacy. *Millennium: Journal of International Studies*, 31(3), 627–51.

Neumann, I. B. (2007). 'A Speech That the Entire Ministry May Stand For', Or: Why Diplomats Never Produce Anything New. *International Political Sociology*, 1, 183–200.

Newell, P. (2000). *Climate for Change: Non-State Actors and the Global Politics of the Greenhouse*. Cambridge: Cambridge University Press.

NOAA. (2012). Trends in Atmospheric Carbon Dioxide. Available at: www.esrl.noaa.gov/gmd/ccgg/trends/ (last accessed 9 August 2012).

Nordlund, G. (2008). Futures Research and the IPCC Assessment Study on the Effects of Climate Change. *Futures*, 40(10), 873–76.

NRC. (2001). *Climate Change Science: An Analysis of Some Key Questions*. Committee on the Science of Climate Change. Washington, DC: National Research Council.

Okem, A. E., Sutherland, C., Roberts, D., Craig, M. H., North, M. A., Hunter, N., & Slotow, R. (2021). 'Reuters' Hot List of Climate Scientists Is Geographically Skewed: Why This Matters'. *The Conversation*. http://theconversation.com/

reuters-hot-list-of-climate-scientists-is-geographically-skewed-why-this-matters-161614 (May 27, 2023).

Okereke, C. (2017). A Six-Component Model for Assessing Procedural Fairness in the Intergovernmental Panel on Climate Change (IPCC). *Climatic Change*, 145(3), 509–22.

Okereke, C., Bulkeley, H., & Schroeder, H. (2009). Conceptualizing Climate Governance beyond the International Regime. *Global Environmental Politics*, 9(1), 58–78.

O'Neill, S., Williams, H. T. P., Kurz, T., Wiersma, B., & Boykoff, M. (2015). Dominant Frames in Legacy and Social Media Coverage of the IPCC Fifth Assessment Report. *Nature Climate Change*, 5(4), 380–85. Article 4.

O'Reilly, J. (2022). Uncertainty. In K. De Pryck & M. Hulme, eds., *A Critical Assessment of the Intergovernmental Panel on Climate Change*. Cambridge: Cambridge University Press, 159–68.

Overland, I., Fossum Sagbakken, H., Isataeva, A., Kolodzinskaia, G., Simpson, N. P., Trisos, C., & Vakulchuk, R. (2022). Funding Flows for Climate Change Research on Africa: Where Do They Come from and Where Do They Go? *Climate and Development*, 14(8), 705–24.

Paasi, A. (2005). Globalisation, Academic Capitalism, and the Uneven Geographies of International Journal Publishing Spaces. *Environment and Planning A: Economy and Space*, 37(5), 769–89.

Pachauri, R. (2008). 'Some Issues Related to the Future of the IPCC'. Available at: www.ipcc.ch/scoping_meeting_ar5/future-ipcc-4-january-2008-2.pdf(lastaccessed 30 August 2012).

Pachauri, R. (2010). 'Statement by the Chair of the Intergovernmental Panel on Climate Change (IPCC) at the Opening Session of the 16th Conference of the Parties, Cancun, Mexico, November 29, 2010'. Available at: unfccc.int/files/meetings/cop_16/media/application/pdf/101129_cop16_oc_rpac.pdf

Parry, M. L. et al. (2007). *Climate Change 2007: Impacts, Adaptation and Vulnerability.* Contribution of Working Group II to the Fourth Assessment Report of the Intergovernmental Panel on Climate Change. Cambridge University Press, Cambridge, United Kingdom and New York, NY, USA.

Parson, E., Agrawala, S., Patt, A, Keohane, R., Mitchell, R., Botcheva, L., Clark, W. DeSombre, E., McCarthy, J., & Shea, E. (1997). Working Group I: The Forms of Assessment. Available at: https://rmitchel.uoregon.edu/sites/rmitchell1.uoregon.edu/files/resume/other_publications/1997-ExplainingFormsOfAssessment.pdf (last accessed 30 May 2023).

Pasgaard, M. & Strange, N. (2013). A Quantitative Analysis of the Causes of the Global Climate Change Research Distribution. *Global Environmental Change*, 23, 1684–93.

Pasgaard, M., Dalsgaard, B., Maruyama, P. K., Sandel, B., & Strange, N. (2015). Geographical Imbalances and Divides in the Scientific Production of Climate Change Knowledge. *Global Environmental Change*, 35, 279–88.

Paterson, M. (1996). *Global Warming and Global Politics*. London: Routledge.

Paterson, M. (2009). Post-Hegemonic Climate Politics? *British Journal of Politics & International Relations*, 11, 140–58.

Paterson, M. & Grubb, M. (1992). The International Politics of Climate Change. *International Affairs*, 68(2), 293–310.

PBL. (2010). Assessing an IPCC Assessment. An Analysis of Statements on Projected Regional Impacts in the 2007 Report. *Netherlands Environmental Assessment Agency (PBL)*. The Hague/Bilthoven.

Pearce, D. W. et al. (1996). The Social Costs of Climate Change: Greenhouse Damage and the Benefits of Control. In J. P. Bruce, H. Yi, & E. F. Haites, eds., *Climate Change 1995: Economic and Social Dimensions of Climate Change: Contribution of Working Group III to the Second Assessment Report of the Intergovernmental Panel on Climate Change*. Cambridge: Cambridge University Press, 179–224.

Pearce, F. (2010). Climate Wars: The Story of the Hacked Emails. *Guardian Special Investigation*, 9 February 2010. Available at: www.guardian.co.uk/environment/series/climate-wars-hacked-emails (last accessed 13 August 2012).

Pearce, W., Mahony, M., & Raman, S. (2018). Science Advice for Global Challenges: Learning from Trade-offs in the IPCC. *Environmental Science & Policy*, 80, 125–31.

Pearson, A. R. & Schuldt, J. P. (2014). Facing the Diversity Crisis in Climate Science. *Nature Climate Change*, 4(12), 1039–42. Article 12.

Petersen, A. C. (2006). *Simulating Nature: A Philosophical Study of Computer-Simulation Uncertainties and Their Role in Climate Science and Policy*. The Netherlands: Het Spinhuis Publishers.

Pouliot, V. (2010). *International Security in Practice: The Politics of Nato-Russia Diplomacy*. Cambridge: Cambridge University Press.

Pouliot, V. (2016). *International Pecking Orders: The Politics and Practice of Multilateral Diplomacy*. Cambridge, UK: Cambridge University Press.

Pouliot, V. & Mérand, F. (2012). Bourdieu's Concepts: Political Sociology in International Relations. In R. Adler-Nissen (ed) *Bourdieu in International Relations*. Routledge: London and New York. 24–44.

Raffles, H. (2014). *In Amazonia: A Natural History*. Princeton and Oxford: Princeton University Press.

Rajamani, L. (2015a). The Devilish Details: Key Legal Issues in the 2015 Climate Negotiations. *The Modern Law Review*, 78(5), 826–53.

Rajamani, L. (2015b). Addressing Loss and Damage from Climate Change Impacts. *Economic and Political Weekly*, 50(30), 17–21.

Rajamani, L. (2016). Ambition and Differentiation in the 2015 Paris Agreement: Interpretive Possibilities and Underlying Politics. *International & Comparative Law Quarterly*, 65(2), 493–514.

Reuters. (2021). Explore the Reuters Hot List of 1,000 Top Climate Scientists. *Reuters*. Apri20, 2021. www.reuters.com/investigates/special-report/climate-change-scientists-list/ (last accessed 2023 May 16).

Riles, A. (2000). *The Network Inside Out*. Michigan: University of Michigan Press.

Rodhe, H. (1991). Bolin Bert and His Scientific Career. *Tellus Series a-Dynamic Meteorology and Oceanography* 43:3–7.

Sagar, A. & Kandlikar, M. (1997). Knowledge, Rhetoric and Power – International Politics of Climate Change. *Economic and Political Weekly*, 31, 39–48.

Schenuit, F. (2023). Staging Science: Dramaturgical Politics of the IPCC's Special Report on 1.5 °C. *Environmental Science & Policy*, 139, 166–76.

Schipper, E. L. F., Ensor, J., Mukherji, A., Mirzabaev, A., Fraser, A., Harvey, B., Totin, E., Garschagen, M., Pathak, M., Antwi-Agyei, P., Tanner, T., & Shawoo, Z. (2021). Equity in Climate Scholarship: A Manifesto for Action. *Climate and Development*, 13(10), 853–56.

Schneider, S. H. (1991). Three Reports of the Intergovernmental Panel on Climate Change. *Environment*, 33, 25–30.

Schneider, S. H. (2009). *Science as a Contact Sport: Inside the Battle to Save Earth's Climate*. Washington, DC: National Geographic.

Schumacher, E. F. (1974). *Small Is Beautiful: A Study of Economics as If People Mattered.* [S.l.]. London: Sphere Books.

SCOPE. (1986). *The Greenhouse Effect, Climatic Change, and Ecosystems.* In B. Bolin & B. R. Doos et al. eds., *SCOPE 29*. Chichester: John Wiley & Sons, 541.

Scott, L., Bucher, A., Cross, A., Kunaratnam, Y., & Tufet, M. (2021). UK-Funded Research on Climate Change and International Development: The Scope and Reach of UK ODA and Wellcome-Funded Research (2015–2020). UK Collaborative on Development Research (UKCDR). Available at: www.ukcdr.org.uk/wp-content/uploads/2021/04/01880-UKCDR-Climate-Change-Report.pdf (last accessed 12 June 2023).

Sending, O. J. (2015). *The Politics of Expertise: Competing for Authority in Global Governance*. Ann Arbor: University of Michigan Press.

Seitz, F. (1996). A Major Deception on Global Warming. *Wall Street Journal*, June 12, 1996.

Shackley, S. (1996). Global Climate Change and Modes of International Science and Policy. In A. Elzinga & C. Langstrom (eds), *Internationalism and Science*. Taylor Graham, London, 199–222.

Shackley, S. (1999). Climate Change Science and Policy in the UK: An Over-Identified Scientific Problem in a Context of Political Intransigence. Available at: http://pne.people.si.umich.edu/PDF/PMNPC/uk.pdf (last accessed 23 September 2012).

Shackley, S. & T. Skodvin. (1995). Ipcc Gazing and the Interpretative Social Sciences: A Comment on Sonja Boehmer- Christiansen's: 'Global Climate Protection Policy: The Limits of Scientific Advice'. *Global Environmental Change* 5:175–80.

Shackley, S. & Wynne, B. (1995). Global Climate Change: The Mutual Construction of an Emergent Science-Policy Domain. *Science and Public Policy*, 22, 218–30.

Shackley, S. & Wynne, B. (1996). Representing Uncertainty in Global Climate Change Science and Policy: Boundary-Ordering Devices and Authority. *Science Technology and Human Values*, 21, 275–302.

Shackley, S., Young, P., Parkinson, S., & Wynne, B. (1998). Uncertainty, Complexity and Concepts of Good Science in Climate Change Modelling: Are GCMs the Best Tools? *Climatic Change*, 38(2), 159–205.

Shaw, A. (2000). *Imbued Meaning: Science-Policy Interactions in the Intergovernmental Panel on Climate Change*. Honours, BA: The University of British Columbia.

Shaw, A. (2005). Policy Relevant Scientific Information: The Co-production of Objectivity and Relevance in the IPCC. Breslauer Symposium, University of California International and Area Studies, UC Berkeley.

Sheirmeier, Q. (2010). IPCC Signs Up for Reform: Panel Agrees New Guidelines and Management Restructure with Pachauri Still at the Helm. *Nature*, 467, 891–92.

Siebenhüner, B. (2002). How Do Scientific Assessments Learn? Part 1. Conceptual Framework and Case Study of the IPCC. *Environmental Science & Policy*, 5(5), 411–20.

Siebenhüner, B. (2003). The Changing Role of Nation States in International Environmental Assessments: The Case of the IPCC. *Global Environmental Change – Human and Policy Dimensions*, 13, 113–23.

Skodvin, T. (2000a). *Structure and Agent in the Scientific Diplomacy of Climate Change: An Empirical Case Study of Science-Policy Interaction in the Intergovernmental Panel on Climate Change*. Advances in Global Change Research. Dordrecht; London: Kluwer Academic.

Skodvin, T. (2000b). Revised Rules of Procedures for the IPCC Process. *Climatic Change*, 46, 409–15.

SMIC. (1971). *Inadvertent Climate Modification: Report of the Study of Man's Impact on Climate (Smic)*. Cambridge, MA; London: M.I.T. Press.

Smith J. B. et al. (2001). *Vulnerability to Climate Change and Reasons for Concern: A Synthesis Climate Change 2001 – Impacts, Adaptation, and Vulnerability*. Contribution of the Working Group II to the Third Assessment Report of the Intergovernmental Panel on Climate Change. J. J. McCarthy, O. F. Canziani, N. A. Leary, D. J. Dokken & K. S. White eds., Cambridge: Cambridge University Press, 913–67.

Solomon, S. (2007). *Climate Change 2007: The Physical Science Basis: Contribution of Working Group I to the Fourth Assessment Report of the Intergovernmental Panel on Climate Change*. Cambridge: Cambridge University Press.

Sorenson, R. P. (2011). Eunice Foote's Pioneering Research on CO2 and Climate Warming. Search and Discovery Search and Discovery, #70092.

Standring, A. (2022). Participant Diversity. In K. De Pryck & M. Hulme eds., *A Critical Assessment of the Intergovernmental Panel on Climate Change*. Cambridge: Cambridge University Press, 61–70.

Standring, A. & Lidskog, R. (2021). (How) Does Diversity Still Matter for the IPCC? Instrumental, Substantive and Co-Productive Logics of Diversity in Global Environmental Assessments. *Climate*, 9(6), 1–15. Article 6.

Stavins, R. (2010). Will We Know Success When We See It? *National Journal,* 6 December 2010. Available at: http://climate.nationaljournal.com/2010/12/will-we-know-success-when-we-s.php (last accessed 13 August 2012).

Stavins, R. (2014). Is the Government Approval Process Broken? Blog Post at An Economic View of the Environment: A Blog by Robert Stavins, posted April 25 2014. Available at: www.robertstavinsblog.org/2014/04/25/is-the-ipcc-%20government-approval-process-broken-2/ (last accessed 25 September 2019)

Stuvøy, K. (2010). Human Security Research Practices: Conceptualizing Security for Women's Crisis Centres in Russia. *Security Dialogue*, 41, 279–99.

Sundqvist, G., Bohlin, I., Hermansen, E. A., & Yearley, S. (2015). Formalization and Separation: A Systematic Basis for Interpreting Approaches to Summarizing Science for Climate Policy. *Social Studies of Science*, 45, 416–40.

Tandon, A. (2021). Analysis: The Lack of Diversity in Climate-Science Research. *Carbon Brief,* posted 5 October 2021. Available at: www.carbonbrief.org/analysis-the-lack-of-diversity-in-climate-science-research/ (last accessed 5 June 2023).

Tandon, A. (2023). Analysis: How the Diversity of IPCC Authors Has Changed over Three Decades. *Carbon Brief,* posted 15 March 2023. Available at: www.carbonbrief.org/analysis-how-the-diversity-of-ipcc-authors-has-changed-over-three-decades/ (last accessed 21 May 2023).

Templeton, J., Davenport, D., Guitiérrez, M., & Hansen, G. (2022). Summary of the 56th Session of the Intergovernmental Panel on Climate Change and the 14th Session of Working Group III: 21 March–4 April 2022. *Earth Negotiations Bulletin*, 12 (795), 1–32.

Tessnow-von Wysocki, I. & Vadrot, A. B. M. (2020). The Voice of Science on Marine Biodiversity Negotiations: A Systematic Literature Review. *Frontiers in Marine Science*, 7, 1–26.

Tschakert, P. (2015). 1.5°C or 2°C: A Conduit's View from the Science-Policy Interface at COP20 in Lima, Peru. *Climate Change Responses*, 2(1), 3.

UNFCCC. (1992). *United Nations Framework Convention on Climate Change*. Geneva: United Nations, United Nations Office at Geneva.

UNFCCC. (1997). *Kyoto Protocol to the United Nations Framework Convention on Climate Change*. Geneva: United Nations, United Nations Office at Geneva.

UNFCCC. (2015). Paris Agreement. FCCC/CP/2015/L.9/Rev.1. United Nations, United Nations Office at Geneva, Geneva.

UNGA RES 45/53. (1988). Protection of Global Climate for Present and Future Generations of Mankind.

UNGA RES 45/212. (1990). Protection of Global Climate for Present and Future Generations of Mankind.

van der Sluijs, J., van Eijndhoven, J., Shackley, S., & Wynne, B. (1998). Anchoring Devices in Science for Policy: The Case of Consensus around Climate Sensitivity. *Social Studies of Science*, 28(2), 291–323.

Vecchione Gonçalves, M. with Hughes, H. (2023). Stakes: Conducting Relational Research with Indigenous Peoples. In H. Hughes & A. B. M. Vadrot, eds., *Conducting Research on Global Environmental Agreement-Making*. Cambridge: Cambridge University Press, 74–89.

Vergano, D. & O'Driscoll P. (2007). 'Scientists, Governments Clash Over Climate Report'. *USA Today,* 4 August 2007. Available at: www.usatoday.com/weather/climate/globalwarming/2007-04-06-global-warming-report_N.htm# (last accessed 14 August 2011).

Victor, D. G., Gerlagh, R., & Baiocchi, G. (2014). Getting Serious about Categorizing Countries. *Science*, 345(6192), 34–36.

Vogler, J. (1995). *The Global Commons: A Regime Analysis*. Chichester: Wiley.

Wacquant, L. (1989). Towards a Reflexive Sociology: A Workshop with Pierre Bourdieu. *Sociological Theory*, 7(1), 26–63.

Wacquant, L. (1998). Pierre Bourdieu. In R. Stones, ed., *Key Sociological Thinkers*. London: Macmillan Press, 25–229.

Wacquant, L. (2002). The Sociological Life of Pierre Bourdieu. *International Sociology*, 17, 549–56.

Wang, Z., Altenburger, R., Backhaus, T., Covaci, A., Diamond, M. L., Grimalt, J. O., Lohmann, R., Schäffer, A., Scheringer, M., Selin, H., Soehl, A., & Suzuki, N. (2021). We need a global science-policy body on chemicals and waste. *Science*, 371(6531), 774–76.

Weart, S. R. (2008). *The Discovery of Global Warming*. Revised and expanded edition. Cambridge, Massachusetts; London: Harvard University Press.

Wible, B. (2014). IPCC Lessons from Berlin. *Science*, 345(6192), 34–34.

Williams, M. C. (2007). *Culture and Security: Symbolic Power and the Politics of International Security*. London: Routledge.

WMO. (1979). Declaration of the World Climate Conference. In Proceedings of the World Climate Conference – A Conference of Experts in Climate and Mankind, 12–23 February 1979, WMO Publication no. 537, Geneva: World Meteorological Organization

WMO. (1986). Report Of The International Conference On The Assessment Of The Role Of Carbon Dioxide And Of Other Greenhouse Gases In Climate Variations And Associated Impacts. Villach, Austria, 1985, WMO/UNEP/ICSU.

World Commission on Environment and Development. (1987). *Our Common Future*. Oxford: Oxford University Press.

Yamin, F. & Depledge, J. (2004). *The International Climate Change Regime: A Guide to Rules, Institutions and Procedures*. Cambridge: Cambridge University Press.

Yamineva, Y. (2010). The Assessment Process of the Intergovernmental Panel on Climate Change (IPCC). Submitted as PhD thesis. Newnham College, Cambridge.

Yamineva, Y. (2017). Lessons from the Intergovernmental Panel on Climate Change on inclusiveness across geographies and stakeholders. *Environmental Science & Policy*, 77(Supplement C), 244–51.

Yearley, S. (2009). Sociology and Climate Change after Kyoto What Roles for Social Science in Understanding Climate Change? *Current Sociology*, 57(3), 389–405.

Zillman, J. W. (2007). Some Observations on the IPCC Assessment Process 1988–2007. *Energy and Environment*, 18, 869–92.

Zillman, J. W. (2008). Australian Participation in the Intergovernmental Panel on Climate Change. *Energy & Environment*, 19, 21–42.

Zommers, Z., Marbaix, P., Fischlin, A., Ibrahim, Z. Z., Grant, S., Magnan, A. K., Pörtner, H.-O., Howden, M., Calvin, K., Warner, K., Thiery, W., Sebesvari, Z., Davin, E. L., Evans, J. P., Rosenzweig, C., O'Neill, B. C., Patwardhan, A., Warren, R., van Aalst, M. K., & Hulbert, M. (2020). Burning Embers: Towards More Transparent and Robust Climate-Change Risk Assessments. *Nature Reviews Earth & Environment*, 1(10), 516–29. Article 10.

Index

For EU product safety concerns, contact us at Calle de José Abascal, 56–1°, 28003 Madrid, Spain or eugpsr@cambridge.org.

www.ingramcontent.com/pod-product-compliance
Lightning Source LLC
LaVergne TN
LVHW060055220926
848484LV00006BA/1364

* 9 7 8 1 0 0 9 3 4 1 5 3 0 *